Praise for

HELGOLAND

by Carlo Rovelli

"Popular science has rarely been so good."

—*Prospect* (UK)

"One of the warmest, most elegant and most lucid interpreters to the laity of the dazzling enigmas of his discipline . . . [A] momentous book."

—John Banville, *The Wall Street Journal*

"Bracing and refreshing . . . Rovelli is offering a new way to understand not just the world but our place in it, too."

—NPR

"A remarkably wide-ranging new meditation on quantum theory with the light touch of a skilled storyteller."

—*The Guardian*

"Rovelli tackles both the quantum realm and the ways it helps us make sense of the mind with refreshing clarity."

—Anil Ananthaswamy, *The New York Times Book Review*

"Explained with uncanny insight and lyrical grace."

—*Time*

"This entertaining and legible guide paints the history of quantum theory and lays out its possible meanings."

—*Scientific American*

"[An] intellectually exhilarating dive into the profoundest scientific conundrums."

—*Booklist* (starred review)

"Physicist Rovelli (*The Order of Time*) dazzles with this look at the 'almost psychedelic experience' of understanding quantum theory. . . . These are big ideas, but Rovelli easily leads readers through the knotty logic, often with lyricism. . . . Readers who follow along will be left in awe."

—*Publishers Weekly* (starred review)

"Charmingly idiosyncratic . . . [Rovelli is] perhaps the finest author now writing on physics and the quantum world."

—*The Article*

"*Helgoland* is Rovelli's most beautiful yet. . . . Unforgettable."

—*The Times* (London)

"Another brilliant book by Rovelli, who is emerging as the most approachable yet authoritative contemporary writer about quantum physics. He describes a 'relational' universe in which there are no absolutes and everything depends on interactions between objects. You won't understand quantum mechanics after reading this—or any other book—but you'll have fun trying."

—*Financial Times*

"Rovelli is a deep-thinking, restlessly inquiring spirit."

—*The Observer* (London)

"Inspiring."

—*The Spectator*

ALSO BY CARLO ROVELLI

The Order of Time

Reality Is Not What It Seems: The Journey to Quantum Gravity

Seven Brief Lessons on Physics

The First Scientist: Anaximander and His Legacy

HELGOLAND

MAKING SENSE OF THE QUANTUM REVOLUTION

Carlo Rovelli

Translated by Erica Segre and Simon Carnell

RIVERHEAD BOOKS · NEW YORK

RIVERHEAD BOOKS
An imprint of Penguin Random House LLC
penguinrandomhouse.com

Originally published in Italy as *Helgoland* by Adelphi Edizioni, Milan, in 2020.
First published in English in Great Britain by Allen Lane, an imprint of
Penguin Random House Ltd., London, in 2021.
First North American edition published by Riverhead Books, 2021.

Page 223 constitutes an extension of this copyright page.

The Library of Congress has catalogued the Riverhead hardcover edition as follows:

Names: Rovelli, Carlo, 1956– author. | Segre, Erica, translator. |
Carnell, Simon, 1962– translator.
Title: Helgoland : making sense of the quantum revolution / Carlo Rovelli ;
translated by Erica Segre and Simon Carnell.
Other titles: Helgoland. English
Description: First U.S. hardcover. | New York : Riverhead Books, 2021. |
"Originally published in Italian under the title Helgoland by Adelphi Edizioni,
Milan, in 2020"—Title page verso. | Includes bibliographical references and index.
Identifiers: LCCN 2020049573 (print) | LCCN 2020049574 (ebook) |
ISBN 9780593328880 (hardcover) | ISBN 9780593328903 (ebook)
Subjects: LCSH: Quantum theory.
Classification: LCC QC173.96 .R6813 2021 (print) |
LCC QC173.96 (ebook) | DDC 530.12—dc23
LC record available at https://lccn.loc.gov/2020049573
LC ebook record available at https://lccn.loc.gov/2020049574

First Riverhead hardcover edition: May 2021
First Riverhead trade paperback edition: May 2022
Riverhead trade paperback ISBN: 9780593328897

Printed in the United States of America
1st Printing

Book design by Amanda Dewey

To Ted Newman,

who made me understand

that I did not understand quantum theory

CONTENTS

HELGOLAND

LOOKING INTO THE ABYSS

Časlav and I are sitting on the sand a few steps from the shore. We have been talking intensely for hours. We came to the island of Lamma, across from Hong Kong, during the afternoon break of a conference. Časlav is a world-renowned expert on quantum mechanics. At the conference, he presented an analysis of a complex thought experiment. We discussed and rediscussed the experiment on the path through the coastal jungle leading to the shore, and then here, by the sea. We have ended up basically agreeing. On the beach there is a long silence. We watch the sea. "It's really incredible," Časlav whispers. "Can this be believed? It's as if reality . . . didn't exist . . ."

This is the stage we are at with quanta. After a century of resounding triumphs, having gifted us contemporary technology and the very basis for twentieth-century physics, the theory that is one of the greatest achievements of science fills us with astonishment, confusion and disbelief.

There was a moment when the grammar of the world seemed clear: at the root of the variegated forms of reality, just particles of matter guided by a few forces. Humankind could think that it had raised the Veil of Maya, seen the basis of the real. It didn't last. Many facts did not fit. Until, in the summer of 1925, a twenty-three-year-old German spent days of anxious solitude on a windswept island in the North Sea: Helgoland—in English also Heligoland—the Sacred Island. There, on the island, he found the idea that made it possible to account for all recalcitrant facts, to build the mathematical structure of quantum mechanics, "quantum theory." Perhaps the most impressive scientific revolution of all time. The name of the young man was Werner Heisenberg, and the story told in this book begins with him.

Quantum theory has clarified the foundations of chemistry, the functioning of atoms, of solids, of plasmas, of the color of the sky, the dynamics of the stars, the origins of galaxies . . . a thousand aspects of the world. It is at the basis of the latest technologies: from computers to nuclear power. Engineers, astrophysicists, cosmologists, chemists and biologists all use it daily; the rudiments of the theory are included in high school curricula. It has never been wrong. It is the beating heart of today's science. Yet it remains profoundly mysterious, subtly disturbing.

It has destroyed the image of reality as made up of particles that move along defined trajectories—without, however, clarifying how we should think of the world instead. Its mathematics does not describe reality. Distant objects seem magically connected. Matter is replaced by ghostly waves of probability.

Whoever stops to ask themselves what quantum theory has to say about the actual world remains perplexed. Albert Einstein, even though he had anticipated ideas that put Heisenberg on the right track, could never digest it himself. Richard Feynman, the great theoretical physicist of the second half of the twentieth century, wrote that nobody understands quanta.

But this is what science is all about: exploring new

ways of conceptualizing the world. At times, radically new. It is the capacity to constantly call our concepts into question. The visionary force of a rebellious, critical spirit, capable of modifying its own conceptual basis, capable of redesigning our world from scratch.

If the strangeness of quantum theory confuses us, it also opens new perspectives with which to understand reality. A reality that is more subtle than the simplistic materialism of particles in space. A reality made up of *relations* rather than objects.

The theory suggests new directions in which to rethink great questions, from the structure of reality to the nature of experience, from metaphysics to perhaps even the very nature of consciousness. Today this is all a matter of the liveliest debate among scientists and among philosophers. I speak about it all in the following pages.

On the island of Helgoland—barren, extreme, battered by the winds of the north—Werner Heisenberg lifted a veil. An abyss opened. The story that this book has to tell starts from the island where Heisenberg conceived the germ of his idea, and progressively widens to take in ever bigger questions opened by the discovery of the quantum structure of reality.

LOOKING INTO THE ABYSS

I have written this book primarily for those who are unfamiliar with quantum physics and are interested in trying to understand, as far as any of us can, what it is and what it implies. I have sought to be as concise as possible, omitting every detail that is not essential to grasping the heart of the issue. I have tried to be as clear as possible, about a theory that is at the center of the obscurity of science. Perhaps rather than explaining how to understand quantum mechanics, I explain why it is so difficult to understand.

But I have also written it thinking of my colleagues—scientists and philosophers, who, the more they delve into the theory, the more they are perplexed—to continue the ongoing conversation on the significance of this astonishing physics. The book has notes intended for those who are familiar with quantum mechanics. They add a bit of precision to what I try to say in a more readable form in the text.

The objective of my research in theoretical physics has been to understand the quantum nature of space and time: to make quantum theory cohere with Einstein's discoveries. For this, I have found myself thinking continually about quanta. This book represents where I have gotten to so far. It does not ignore other opinions, but it is shamelessly partisan: centered on the perspective that I consider the most effective and that I think opens up the most interesting paths: the "relational" interpretation of quantum theory.

A warning before we begin. The abyss of what we do not know is always magnetic and vertiginous. But to take quantum mechanics seriously, reflecting on its implications, is an almost psychedelic experience: it asks us to renounce, in one way or another, something that we cherished as solid and untouchable in our understanding of the world. We are asked to accept that reality may be profoundly other than we had imagined: to look into the abyss, without fear of sinking into the unfathomable.

—Lisbon, Marseille, Verona,
and London, Ontario
2019–20

PART ONE

I

A STRANGELY BEAUTIFUL INTERIOR

How a young German physicist arrived at an idea that was very strange indeed, but described the world remarkably well—and the great confusion that followed.

THE ABSURD IDEA OF THE YOUNG HEISENBERG: OBSERVABLES

It was around three o'clock in the morning when the final results of my calculations were before me. I felt profoundly shaken. I was so agitated that I could not sleep. I left the house and began walking slowly in the dark. I climbed on a rock overlooking the sea at the tip of the island, and waited for the sun to come up . . .[1]

I have often wondered what the thoughts and emotions of the young Heisenberg must have been as he clambered over that rock overlooking the sea, on the barren and windswept North Sea island of Helgoland, facing the vastness of the waves and awaiting the sunrise, after having been the first to glimpse one of the most vertiginous of Nature's secrets ever looked upon by humankind. He was twenty-three.

He was on the island seeking relief from the allergy that afflicted him. Helgoland—the name means Sacred Island—has virtually no trees, and very little pollen. ("Heligoland with its one tree," as James Joyce has it in *Ulysses*.) Perhaps the legends of the dreadful pirate Störtebeker hiding on the island, which Heisenberg loved as a boy, were in his mind as well. But Heisenberg's main reason for being there was to immerse himself in the problem with which he was obsessed, the burning issue handed to him by Niels Bohr. He slept little and spent his time in solitude, trying to calculate something that would justify Bohr's incomprehensible rules. Every so often, he would take a break to climb over the island's rocks or learn by heart poetry from Goethe's *West-Eastern Divan*, the collection in which Germany's greatest poet sings his love for Islam.

Niels Bohr was already a renowned scientist. He had written formulas, simple but strange, that predicted the properties of chemical elements even before measuring them. They predicted, for instance, the frequency of light emitted by elements when heated: the color they assume. This was a remarkable achievement. The formulas, however, were incomplete: they did not give, for instance, the intensity of the emitted light.

But above all, these formulas had about them something that was truly absurd. They assumed, for no good reason, that the electrons in atoms orbited around the nucleus only on *certain* precise orbits, at *certain* precise distances from the nucleus, with *certain* precise energies—before magically "leaping" from one orbit to another. The first quantum leaps. Why only these orbits? Why these incongruous "leaps" from one orbit to another? What force could possibly cause such bizarre behavior as this?

The atom is the building block of everything. How does it work? How do the electrons move inside it? The scientists of the beginning of the century had been pondering these questions for more than a decade, without getting anywhere.

Like a Renaissance master painter in his studio, Bohr

had gathered around him in Copenhagen the very best young physicists he could find, to work together on the mysteries of the atom. Among them was the brilliant Wolfgang Pauli—Heisenberg's extremely intelligent, pretty arrogant friend and former classmate. But Pauli had recommended Heisenberg to the great Bohr, saying that to make any real progress, he was needed. Bohr had taken the advice, and in the autumn of 1924 had brought Heisenberg to Copenhagen from Göttingen, where he was working as an assistant to the physicist Max Born. Heisenberg had spent a few months in long discussions with Bohr, in Copenhagen, in front of blackboards covered with formulas. The young apprentice and the master had taken long walks together in the mountains, talking about the enigmas of the atom; about physics and philosophy.[2]

Heisenberg had steeped himself in the problem. It had become his obsession. Like the others, he had tried everything. Nothing worked. There seemed to be no reasonable force capable of guiding the electrons on Bohr's strange orbits, and in his peculiar leaps. And yet those orbits and those leaps really did lead to good predictions of atomic phenomena. Confusion.

Desperation pushes us to look for extreme solutions.

On that island in the North Sea, in complete solitude, Heisenberg resolved to explore radical ideas.

It was with radical ideas, after all, that twenty years earlier Einstein had astonished the world. Einstein's radicalism had worked. Pauli and Heisenberg were enamored of his physics. Einstein for them was a legend. Had the time perhaps come, they asked themselves, to hazard as radical a step, to escape from the impasse regarding electrons in atoms? Could they be the ones to take it? In your twenties, you can dream freely.

Einstein had shown that even our most rooted convictions can be wrong. What seems most obvious to us now might turn out not to be correct. Abandoning assumptions that seem self-evident can lead to greater understanding. Einstein had taught that everything should be based on what we see, not on what we assume to exist.

Pauli repeated these ideas to Heisenberg. The two young men had drunk deep of this poisoned honey. They had been following the discussions on the relation between reality and experience that ran through Austrian and German philosophy at the beginning of the century. Ernst Mach, who had exerted a decisive influence on Einstein, insisted that knowledge had to be based solely on observations, freed of any implicit "metaphysical" assumption. These were the ingredients coming together in the young Heisenberg's thinking, like the chemical components of an explosive, as he isolated himself on Helgoland in the summer of 1925.

And here he had the idea. An idea that could only be had with the unfettered radicalism of the young. The idea that would transform physics in its entirety—together with the whole of science and our very conception of the world. An idea, I believe, that humanity has not yet fully absorbed.

Heisenberg's leap is as daring as it is simple. No one has been able to find the force capable of causing the bizarre behavior of electrons? Fine, let's stop searching for this new force. Let's use instead the force we are familiar with: the electric force that binds the electron to the nucleus. We cannot find new laws of motion to account for Bohr's orbits and his "leaps"? Fine, let's stick with the laws of motion that we're familiar with, without altering them.

Let's change, instead, our way of thinking about the electron. Let's give up describing its movement. Let's describe *only what we can observe*: the light it emits. Let's base everything on quantities that are *observable*. This is the idea.

Heisenberg attempts to recalculate the behavior of the electron using quantities we observe: the frequency and amplitude of emitted light.

We can observe the effects of the electron's *leaps* from one of Bohr's orbits to another. Heisenberg replaces the physical variables (numbers) with *tables of numbers* that have the orbits of departure in their rows and the orbits of arrival in their columns. Each entry of the table stands in a row and in a column: it describes the leap from one orbit to another. He spends his time on the island trying to use these tables to calculate something that could justify Bohr's rules. He doesn't get much sleep. But he fails to do the math for the electron in the atom: too difficult. He tries to account for a simpler system instead, choosing a pendulum, and looks for Bohr's rules in this simpler case.

On June 7, something begins to click:

> When the first terms seemed to come right [giving Bohr's rules], I became excited, making one mathematical error after another. As a consequence, it was around three o'clock in the morning

> when the result of my calculations lay before me. It was correct in all terms.
>
> Suddenly I no longer had any doubts about the consistency of the new "quantum" mechanics that my calculation described.
>
> At first, I was deeply alarmed. I had the feeling that I had gone beyond the surface of things and was beginning to see a strangely beautiful interior, and felt dizzy at the thought that now I had to investigate this wealth of mathematical structures that Nature had so generously spread out before me.

It takes our breath away. Beyond the surface of things, "a strangely beautiful interior." Heisenberg's words resonate with those written by Galileo on first seeing the mathematical regularity appear in his measurements of the fall of objects along an inclined plane: the first mathematical law describing the motion of objects on Earth ever discovered by humankind. Nothing is like the emotion of seeing a mathematical law behind the disorder of appearances.

ሐ

On June 9, Heisenberg leaves Helgoland and returns to his university in Göttingen. He sends a copy of his results to his friend Pauli, with the comment "Everything is still very vague and unclear to me, but it seems that electrons no longer move in orbits."

On July 9, he sends a copy of his work to Max Born, the professor he was assisting, with a note saying: "I have written a crazy paper and do not have the courage to submit it anywhere for publication." He asks Born to read it and to advise.

On July 25, Max Born himself sends Heisenberg's work to the scientific journal *Zeitschrift für Physik*.[3]

Born has seen the importance of the step taken by his young assistant. He seeks to clarify matters. He gets his student Pascual Jordan involved in trying to bring order to Heisenberg's outlandish results.[4] For his part, Heisenberg tries to get Pauli involved, but Pauli is unconvinced: it all seems to him like a mathematical game, far too abstract and abstruse. At first it is just the three of them working on the theory: Heisenberg, Born and Jordan.

They work feverishly, and in just a few months manage to put in place the entire formal structure of a new mechanics. It is very simple: the forces are the same as in classical physics; the equations are the same as those of classical physics (plus one,* which I will talk about later). But the variables are replaced by tables of numbers, or "matrices."

ሰሰ

Why tables of numbers? What we observe of an electron in an atom is the light emitted when, according to Bohr's hypothesis, it leaps from one orbit to another. A leap involves *two* orbits: the one the electron leaves and the one it leaps to. Each observation can then be placed, as I have mentioned, in the entries of a table where the orbit of departure determines the row; the orbit of arrival, the column.

Heisenberg's idea is to write *all* the quantities which describe the movement of the electron—position, velocity, energy—no longer as numbers, but as tables of numbers. Instead of having a single position x for the electron, we have an entire table of possible positions X: one for every possible leap. The idea is to continue to use the *same* equations as always, simply replacing the usual quantities (position, velocity, energy and frequency of orbit and so on) with such tables. Intensity and fre-

* $XP - PX = i\hbar$

quency of light emitted in a leap, for example, will be determined by the corresponding box in the table. The table corresponding to energy has numbers only on the diagonal, and these will give the energies of the Bohr orbits.

Is that clear? It is not. It's as clear as tar.

ORBIT OF DEPARTURE	ORBIT OF ARRIVAL				
	Orbit 1	*Orbit 2*	*Orbit 3*	*Orbit 4*	...
Orbit 1	X_{11}	X_{12}	X_{13}	X_{14}	...
Orbit 2	X_{21}	X_{22}	X_{23}	X_{24}	...
Orbit 3	X_{31}	X_{32}	X_{33}	X_{34}	...
Orbit 4	X_{41}	X_{42}	X_{43}	X_{44}	...
...	...	...	...	...	...

A Heisenberg matrix: the table of numbers that "represent" the position of the electron. The number X_{23}, for example, refers to the leap from the second to the third orbit.

And yet this absurd maneuver of substituting variables with tables enables us to compute the correct results, predicting what is observed in experiments.

To the astonishment of the three Göttingen musketeers, before the year is out, Born receives by post a brief essay by a young Englishman in which essentially the same theory as their own is constructed, using a mathematical language even more abstract than the Göttingen matrices.[5] Its author is Paul Dirac. In June, Heisenberg had given a lecture in England, at the end of which he had mentioned his ideas about quantum leaps. Dirac was in the audience. But he was tired and understood nothing. Later he had been given Heisenberg's first paper by his professor, who had received it by post and found it inscrutable. Dirac reads it, decides it is nonsensical, puts it aside. But a couple of weeks later, reflecting on it during a walk in the countryside, he realizes that Heisenberg's tables resemble something that he has studied in one of his courses. Not remembering what exactly, he has to wait until Monday for the library to open so he can refresh his memory about the ideas in a certain book.[6] From there, in brief, he independently constructs the same complete theory as the three wizards of Göttingen.

All that remains to do is to apply the new theory to the electron in the atom and see if it really works. Will it actually yield all of Bohr's orbits?

The calculation turns out to be difficult, and the three cannot manage to complete it. They seek help from Pauli, the most brilliant as well as the most arrogant of them all. "This is indeed a calculation that is too difficult," he quips, ". . . for you."[7] He completes it, with acrobatic technicality, in the space of a few weeks.[8]

The result is perfect. The energy values calculated using the matrices of Heisenberg, Born and Jordan are precisely those hypothesized by Bohr. Bohr's strange rules for atoms follow from the new scheme. But this is not all. The theory also permits us to compute the intensity of emitted light, as Bohr couldn't. And these also turn out to accord precisely with those obtained in experiments!

It is a complete triumph.

Einstein writes, in a letter to Born's wife, Hedi: "The ideas of Heisenberg and Born have everyone in suspense, and are preoccupying everyone with the slightest interest in theory."[9] And in a letter to his old friend Michele Besso: "The most interesting theorization of recent times is that of Heisenberg-Born-Jordan on quantum states: a calculation of real witchery."[10]

Bohr, the master, will recall years later: "We had at the time only a vague hope of [being able to arrive at] a reformulation of the theory in which every inappropriate use of classical ideas would be gradually eliminated. Daunted by the difficulty of such a program, we all felt great admiration for Heisenberg when, at just twenty-three, he managed it in one swoop."[11]

Except for Born, who is in his forties, Heisenberg, Jordan, Dirac and Pauli are all twentysomethings. In Göttingen they call their physics *Knabenphysik*, or "boys' physics."

ꝏ

Sixteen years later, Europe is in the throes of another world war. Heisenberg is by now a famous scientist. Hitler has assigned to him the task of using his knowledge of the atom to construct a bomb that will win the war. Heisenberg takes a train to Copenhagen, in a Denmark occupied by the German army, and visits his old teacher. The old master and the young man talk together, before parting without having understood each other. Heisenberg will later say that he sought out Bohr to discuss the moral problem entailed by the prospect of a terrifying weapon. Not everyone will believe him. Shortly afterward, with his consent, Bohr is kidnapped in a British commando raid and taken out of occupied Denmark. He is taken to England and received personally by Churchill—then to the United States, where his knowledge is put to work with the generation of young physicists who have learned how to use the new quantum theory to manipulate atoms. Hiroshima and Nagasaki are annihilated. Two hundred thousand human beings—men, women and children—killed in a fraction of a second. Today we live with tens of thousands of nuclear warheads aimed at our cities. If anyone were to lose their head, or make a mistake, there is ample capacity to destroy life on our planet. The devastating power of the "boys' physics" is evident for all to see.

Thankfully, there is much more than weapons. Quantum theory has been applied to atoms, atomic nuclei, elementary particles, the physics of chemical bonds, the physics of solid materials, of liquid and gas, semiconductors, lasers, the physics of stars such as the Sun, neutron stars, the primordial universe, the physics of the formation of galaxies . . . and so on and so forth. The list could go on for pages. Quantum theory has allowed us to understand whole areas of nature, from the form of the periodic table of elements to medical applications that have saved millions of lives. It has predicted new phenomena never previously imagined: quantum correlations over a distance of kilometers, quantum computers, teleportation. All predictions have turned out to be correct. The astonishing run of quantum theory's successes has been uninterrupted for a century, and it continues today.

The calculation scheme by Heisenberg, Born, Jordan and Dirac, the strange idea of "limiting yourself to only what's observable," and to substituting physical variables with matrices,[12] has never yet been wrong. It is the only fundamental theory about the world that until now has never been found wrong—and whose limits we still do not know.

But why is it that we are not able to describe where the electron is and what it is doing when we are *not* observing it? Why must we speak only of its "observables"? Why is it that we can speak of its effect when it leaps from one orbit to another, and yet we cannot say where it is at any given moment? What does it mean to replace *numbers* with *tables* of numbers?

What does it mean that "everything is still very vague and unclear to me, but it seems that electrons no longer move in orbits"? His friend Pauli wrote of Heisenberg: "He reasoned in a terrible way, he was all about intuition; he did not pay any attention to elaborating clearly the fundamental assumptions and their relation to existing theories."

The spellbinding article by Werner Heisenberg, with which everything started, conceived on the island in the North Sea, opens with the phrase: "The objective of this work is to lay the foundations for a theory of quantum mechanics based exclusively on relations between quantities that are in principle observable."

Observable? What does Nature care whether there is anyone to observe or not?

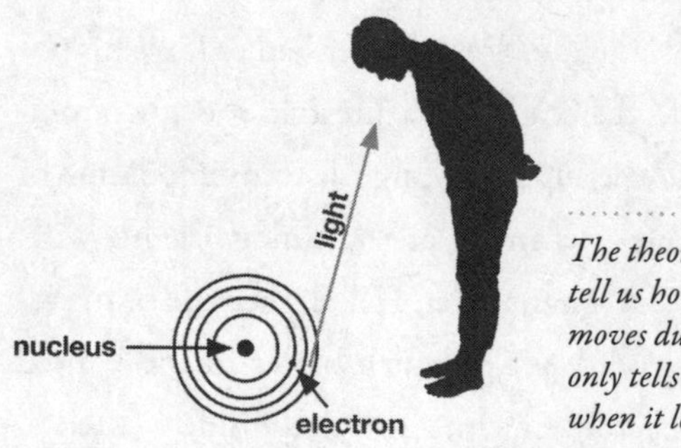

The theory does not tell us how the electron moves during a leap. It only tells us what we see when it leaps. Why?

THE MISLEADING Ψ OF ERWIN SCHRÖDINGER: PROBABILITY

In the following year, 1926, everything seems to come clear. The Austrian physicist Erwin Schrödinger manages to obtain the same result as Pauli, calculating the Bohr energies of the atom, but in a completely different way.

Curiously, this result, too, is not obtained in a university department or lab: Schrödinger achieves it during a getaway with a secret lover in the Swiss Alps. Raised in the libertarian and permissive atmosphere of Vienna at the beginning of the century, the brilliant and charismatic Schrödinger always kept a number of relationships going at once—and made no secret of his fascination with preadolescent girls. Years later, despite being a Nobel laureate, Schrödinger had to leave his position at Oxford because of a lifestyle too unconventional even for the supposed English accommodation of eccentricity. He was living at the time with his wife, Anny, and his pregnant lover, Hilde, the wife of his assistant. Things did not go much better in the United States: at Princeton, Erwin, Anny and Hilde wanted to live together with little Ruth, who had been born in the meantime—but the Ivy League was not ready for a *ménage* such as this. In search of somewhere more liberal, they moved to Dublin—but there, too, Schrödinger ended up at the center of a scandal, after fathering children with two of his students. His wife commented: "You know it would be easier to live with a canary bird than with a race horse, but I prefer the race horse."[13]

The identity of whoever accompanied him into the mountains in those first days of 1926 remains a mystery. We only know that she was an old Viennese friend. Legend has it that he'd headed there taking just his lover, two pearls to place in his ears to isolate himself when he wanted to think about physics, and the thesis of a young French scientist, Louis de Broglie, which Einstein had advised him to read. De Broglie's thesis examines the idea that particles such as electrons might actually be waves—like the waves of the sea, or electromagnetic waves. On the basis of some in fact rather vague theoretical analogies, de Broglie suggests it is possible to think of an electron as being like a small wave in motion.

What kind of relation can there be between a wave, which spreads out, and a particle that remains compact, following a fixed, defined trajectory? Think of a beam of light from a laser: it seems to follow a neat trajectory, like a beam of particles. But it is made of light, which is a wave, an oscillation of the electromagnetic field. The precise line described by the trajectory of a ray of light is an approximation, concealing its dispersion and spread.

Schrödinger is captivated by the idea that the trajectories of elementary particles are also approximations of the behavior of an underlying wave.[14] He had spoken of this idea in a seminar in Zurich, and a student had asked if these waves obeyed an equation. In the mountains, with pearls in his ears, in the intervals between romantic moments shared with his Viennese friend, Schrödinger skillfully works back along the path that leads from the equation of a wave to the trajectory of a ray of light,[15] and in this acrobatic way figures out the equation that the electron-wave must satisfy when it is in an atom. He studies solutions of this equation and manages to extract precisely the Bohr energies.[16] Wow!

Then, having learned about the theory of Heisenberg, Born and Jordan, he succeeds in demonstrating that, from a mathematical point of view, the two theories are substantially equivalent: they predict the same values.[17]

ꕥ

The idea of a wave is so simple that it throws off balance the Göttingen group and their esoteric speculations on observable quantities. It's like Columbus's egg: Heisenberg, Born, Jordan and Dirac had built an intricate and obscure theory only because they had taken the long and winding road. Things are actually simpler: the electron is a wave. That's all. "Observations" have nothing to do with it.

Schrödinger, too, is a product of that lively early-twentieth-century Viennese philosophical and intellectual milieu: a friend of the philosopher Hans Reichenbach, he is fascinated by Asian thought, in particular Vedanta Hinduism, and passionate (as Einstein was) about the philosophy of Arthur Schopenhauer, which interprets the world as "representation." Unconstrained by conformism, having no fear of conventional opinion, of "what people will think," he has no qualms about replacing a world of solid matter with a world of waves.

In naming his wave, Schrödinger uses the Greek letter psi: ψ. The quantity ψ is also called the "wave function."[18] His fabulous calculation seems to show clearly that the microscopic world is not made up of particles: it is made up of ψ waves. Around the nuclei of atoms there are not orbiting specks of matter but the continuous undulation of Schrödinger's waves, like the waves that ruffle the surface of a small lake as the wind blows.

This "wave mechanics" instantly seems more convincing than Göttingen's "matrix mechanics," even if the two give the same predictions. Schrödinger's calculation is simpler than Pauli's. Physicists of the first half of the twentieth century were familiar with waves and wave equations; they were unfamiliar with matrices, the math we call today "linear algebra." A well-known physicist remarked at the time: "Schrödinger's theory came as a relief: we no longer had to learn the peculiar mathematics of matrices."[19]

Above all, Schrödinger's waves are easy to imagine and visualize. They show what it was about the "trajectory of the electron" that Heisenberg wanted to disappear: the electron is a wave that spreads, and that is all. This is why it has no trajectory.

Schrödinger seems to have triumphed in every way.

But it is an illusion.

Heisenberg immediately sees that the conceptual

clarity of Schrödinger's waves is a mirage. Sooner or later a wave spreads out and is spatially diffused; an electron does not: when it arrives from somewhere, it always arrives complete in a single point. If an electron is expelled from a nucleus, Schrödinger's equation predicts that the ψ wave spreads out evenly through space. But when the electron is revealed, by a detector for instance, or by a TV screen, it arrives at one point only, not spatially spread out.

The discussion of Schrödinger's wave mechanics quickly becomes lively, and then suddenly virulent. Heisenberg, who feels that the importance of his discovery is being called into question, is cutting: "The more I think about the physical portion of the Schrödinger theory, the more repulsive I find it. What Schrödinger writes about the visualizability of his theory, 'is probably not right,' in other words it's crap."[20] Schrödinger tries to retort wittily: "I can't imagine that an electron hops about like a flea."[21]

But it is Heisenberg who is right. It becomes gradually evident that *wave mechanics* is no clearer than the *matrix mechanics* of Göttingen. It is another calculation tool that produces the right numbers, and is sometimes easier to use, but it does not give a clear and immediate picture of what happens, as Schrödinger had hoped it would. Wave mechanics is as obscure as the matrices of Heisenberg. If, every time we see an electron, we see it at a single point, how can the electron be a wave diffused in space?

Years later, Schrödinger, who will nevertheless become one of the most acute thinkers on questions about quanta, recognized his defeat. "There was a moment," he writes, "when the creators of wave mechanics [that is, himself; who else?] nurtured the illusion of having eliminated the discontinuities in quantum theory. But the discontinuities eliminated from the equations of the theory reappear the moment the theory is confronted with what we observed."[22]

Once again, we are back with what can be "observed." And once again, this raises the question: What does Nature know or care about whether we are observing it or not?

ꝉꝉ

It is Max Born—him again—who understands for the first time the significance of Schrödinger's ψ, adding a crucial ingredient to the understanding of quantum physics.[23] Born, with his air of a serious but somewhat superannuated engineer, is the least flamboyant and the least well known of the creators of quantum mechanics, but he is perhaps the real architect of the theory—in addition to being, as they say, "the only adult in the room," in an almost literal sense. It was he who in 1925 was clear that quantum phenomena made a radically new mechanics necessary; it was he who had instilled this idea in younger physicists. It was Born, too, who recognized at once the right idea in Heisenberg's first confusing calculation, translating it into a true theory.

What Born understands is that the value of Schrödinger's ψ wave at a point in space is related to the *probability* of observing the electron at this point.[24] If an atom emits an electron and is surrounded by particle detectors, the value of ψ where there is a detector determines the probability of that detector, and not another, detecting the electron.

Schrödinger's ψ is therefore not a representation of a real entity: it is an instrument of calculation that gives the probability that something real will occur. It is like the weather forecasts telling us what could happen tomorrow.

The same—it soon becomes clear—is true of Göttingen matrix mechanics: the mathematics gives predictions that are *probabilistic*, not exact. Quantum theory, just as much in Heisenberg's version as in Schrödinger's, predicts probability, and not certainty.

ꚍꚍ

But why probability? We usually speak of probability when we do not have all the data. The probability that the ball will land on the number five when we spin a roulette wheel is one in thirty-seven. If we knew the exact position of the ball when it was thrown, and all the forces acting on it, we would be able to predict the number on which it would land. (In the 1980s a group of brilliant young gamblers went on a winning streak in a Las Vegas casino, using a small computer concealed in a shoe, exploiting this fact.[25]) When we don't have all of the data, we do not know for certain what will happen—and we speak of probability.

Does this mean that the quantum mechanics of Heisenberg and Schrödinger fails to take all the relevant givens of the problem into account? Is this why we have probability? Or is it that Nature actually leaps about here and there by chance?

Einstein put the question in colorful language: Does God play dice?

Einstein relished figurative language and had a predilection for using "God" in his metaphors despite his declared atheism. But in this case his phrase can be taken literally: he loved Spinoza, for whom "God" was synonymous with "Nature." Hence: "Does God play dice?" means literally: "Are the laws of nature really not deterministic?" As we shall see, a hundred years after Heisenberg and Schrödinger's bickering, this question is still open.

ꕤ

In any case, Schrödinger's ψ wave is definitely not sufficient to clarify the obscurity of the quanta. It is not enough to think of the electron as a wave. The ψ wave is something unclear, which determines the probability that the electron will be observed in one place rather than in another. It evolves in time according to the equation written by Schrödinger, *as long as we do not look at it*. When we look at it, *pffft!*, it disappears, concentrated into a point, and we see the particle there.[26]

As if the mere fact of observing it was enough to modify reality.

To Heisenberg's obscure idea that the theory only describes *observations*, and not what happens between one observation and another, we must add the idea that the theory predicts only the *probability* of observing one thing or another. The mystery deepens.

THE GRANULARITY OF THE WORLD: QUANTA

I have told the story of how quantum theory was born between 1925 and 1926, and have introduced two ideas: the peculiar idea, found by Heisenberg, of describing only *observables*, and the fact that the theory predicts only *probabilities*, understood by Born.

There is a third idea at the core of quantum physics. In order to illustrate it, we had better go back a bit, to the two decades before Heisenberg's fateful journey to the Sacred Island.

The weird behavior of electrons in atoms was not the only strange and incomprehensible phenomenon at the beginning of the twentieth century. Others had also been observed. They had one thing in common: they highlighted a curious *granularity* of energy and other physical quantities. Before the quanta, nobody suspected that energy could be granular. The energy of a thrown stone depends on the velocity of the stone: the velocity can be anything, hence the energy can be anything as well. But peculiar energy behaviors had been appearing in experiments.

⁂

Inside an oven, for example, electromagnetic waves behave in a curious way. The heat (that is, the energy) is not distributed among all frequencies, as we would naturally expect: it never reaches higher frequencies. In 1900, twenty-five years before Heisenberg's journey to Helgoland, the German physicist Max Planck had discovered a formula that reproduced well the way that the energy of heat, measured in a laboratory, distributed among waves of different frequency.[27] He derived this formula from general rules, but added a curious hypothesis: the energy could be transmitted to the waves *only in integer multiples of elementary energies*. In discrete packets.

The dimensions of these packets, in order to make Planck's calculation work, must be different for waves of different frequency: they must be proportional to the frequency.[28] High-frequency waves can only receive more energetic packets. Energy does not reach the very high frequencies because there is not enough of it to transmit big enough packets.

Using experimental observations, Planck had calculated the proportionality constant between the energy of the packets and the frequency. He had called this constant *h*—without understanding its significance. Today we often use the symbol $\hbar$, which stands for *h* divided by 2π. Dirac developed the habit of placing a small line through the *h*, because in calculations *h* is frequently divided by 2π, and he got tired of writing "$h/2\pi$" every time. The symbol $\hbar$ is called "h bar." It has come to be called "Planck's constant," just like *h* without the bar, generating a certain amount of confusion. Today it has become the symbol most characteristic of quantum theory. (I have a T-shirt with a small $\hbar$ embroidered on it, of which I am very fond.)

ꝥꝥ

Five years later, Einstein suggests that light and all the other electromagnetic waves are *actually* made up of elementary grains.[29] These are the first "quanta." Today we call them "photons," the quanta of light. Planck's constant *h* measures their size: every photon has an energy *h* times the frequency of the light of which it is part.

Assuming that these "elementary grains of energy" actually exist, Einstein manages to explain a phenomenon that is not yet understood, called the "photoelectric effect,"[30] and predicts its characteristics before they are measured.

Einstein has provided the inspiration for quantum mechanics in numerous ways. He begins to realize, already in 1905, that the issues raised by these phenomena were serious enough to require a complete revision of mechanics. Born learns from him the idea that mechanics needs to be revised in depth. Einstein's idea that light is a wave but also a cloud of photons inspires de Broglie to think that *all* the elementary particles could be waves, and that leads Schrödinger to introduce the ψ wave. Heisenberg is inspired by Einstein to restrict his attention to quantities that are measurable. There is more: Einstein is also the first to study atomic phenomena using probability, opening the path that leads Born to understand that the meaning of the ψ wave is to be found in probability. Quantum physics owes much to Einstein.

tth

Planck's constant reappears in 1913: in Bohr's rules.[31] Here, too, the same logic: the orbit of the electron in the atom can only have certain energies, as if energy was in packets, granular. When an electron leaps from one of Bohr's orbits to another, it frees a packet of energy that becomes a photon, a quantum of light. Then again, in 1922, an experiment conceived by Otto Stern and carried out by Walter Gerlach in Frankfurt shows that even the rotation speed of atoms is not continuous but takes only certain discrete values.

These phenomena—photons, the photoelectric effect, the distribution of energy among electromagnetic waves, Bohr's orbits, the discreteness of rotation—are all regulated by the Planck constant $\hbar$.

When the quantum theory of Heisenberg and his friends finally appears in 1925, it allows for *all* of these phenomena to be accounted for at a stroke: to predict them, to calculate their characteristics.

The name "quantum theory" comes, indeed, from "quanta," which is to say "grains." "Quantum" phenomena reveal the granular aspect of the world, at a very small scale.

My field of study, quantum gravity, shows that the very physical space in which we live can be granular at a very small scale. The Planck constant determines the (extremely small) scale of the elementary "quanta of space."

Granularity is the third idea of quantum theory, next to *probability* and *observations*. The rows and columns of Heisenberg's matrices correspond directly to the individual *discrete* values that the energy can take.

⁂

We are nearing the end of the first part of the book, the story of the birth of the theory and the confusion it has generated. In the second part, I describe ways out of this confusion.

Before concluding, here are a few words about the single equation that quantum theory adds to classical physics. It is a strange equation. It states that multiplying the position by the velocity is different from multiplying the velocity by the position. If position and velocity were numbers, there could be no difference, because 7 x 9 is the same as 9 x 7. But position and velocity are now *tables* of numbers, and when you multiply two tables, the order counts. The new equation gives us the difference between multiplying two quantities in an order and in the reverse order.

It is beautifully compact, very simple. Incomprehensible.

Do not try to decipher it: scientists and philosophers

are still wrestling with its meaning—and among themselves. Later on, I will return to it to discuss its content a little better. I'll write it now, anyway, because it is the heart of quantum theory. Here it is:

$$XP - PX = i\hbar$$

That's it. The letter X indicates the position of a particle, the letter P indicates its speed multiplied by its mass (what we call "momentum"). The letter i is the mathematical symbol of the square root of -1 and, as we have seen, $\hbar$ is Planck's constant divided by 2π.

In a sense, Heisenberg and company have added to physics *only* this simple equation: everything else follows from it—from the quantum computer to the atomic bomb.

The price of this formal simplicity is the obscurity of its meaning. Quantum theory predicts granularity, quantum leaps, photons and all the rest, on the basis of adding a single equation of eight characters to classical physics. An equation that says to multiply position by speed is different from multiplying speed by position. The opaqueness is complete. Perhaps it is no coincidence that F. W. Murnau is said to have shot some scenes of his classic silent Gothic movie *Nosferatu* on Helgoland.

⁂

In 1927, Niels Bohr gives a lecture at Lake Como, in Italy, in which he summarizes everything that was understood (or not) about the new quantum theory and explains how to use it.[32] In 1930, Dirac writes a book in which the formal structure of the new theory is beautifully elucidated.[33] It is still today the best book to learn it from. Two years later, the greatest mathematician of the time, John von Neumann, puts right the formal issues in a tremendous work of mathematical physics.[34]

The construction of the theory is rewarded by an unparalleled shower of Nobel Prizes. Einstein gets the Nobel in 1921 for having clarified the photoelectric effect by introducing quanta of light; Bohr in 1922 for the rules on the structure of the atom. In 1929, de Broglie is awarded the prize for the idea of waves of matter. In 1932, it is Heisenberg, "for the creation of quantum mechanics," and in 1933, Schrödinger and Dirac, for "new discoveries" in atomic theory. Pauli receives the prize in

1945 for technical contributions to the theory. Born receives it in 1954 for having understood the role of probability (he did so much more besides). The only one to be overlooked is Pascual Jordan, despite the fact that Einstein had (correctly) nominated Heisenberg, Born and Jordan as the true originators of the theory. But Jordan had shown too much loyalty to Nazi Germany, and those defeated in war do not receive credit.[35]

Remaining faithful to Werner Heisenberg's seminal insight on Helgoland, the theory doesn't tell us where to find any one particle of matter when we are not looking at it. It only speaks about the probability of finding it at one point *if we observe it*.

But what does a particle care if we are observing it or not? The most effective and powerful scientific theory is an enigma.

PART TWO

II

A CURIOUS BESTIARY OF EXTREME IDEAS

In which strange quantum phenomena
are described,
and various scientists and philosophers
seek to understand them,
each in their own way.

SUPERPOSITIONS

I hesitated a lot before deciding which academic subject to pursue. I opted for physics at the very last moment. On the day of enrollment at the University of Bologna (it was not yet possible to enroll online), there was a series of queues of different lengths for different programs, and I was helped in my decision by the fact that the physics queue was the shortest.

What attracted me to physics was the suspicion that

beyond the deadly boredom of the subject taught in high school, behind the stupidity of all those exercises with springs, levers and rolling balls, there was a genuine curiosity to understand the nature of reality. A curiosity that resonated with my own restless, adolescent desire to try everything: to read, to know and to see everything, to travel everywhere—to experience as many countries, environments, girls, books, as much music and as many ideas as possible . . .

Adolescence is the time when the networks of neurons in the brain suddenly rearrange themselves. Everything seems intense, everything is alluring, everything is disorienting. I had come out of it confused and full of questions. I wanted to understand the nature of things. I wanted to know how our thinking manages to comprehend this nature. What is reality? What is *thinking*? What is this "I" that does the thinking?

It was this extreme, intense adolescent curiosity that pushed me to seek out what lights science—the great New Knowledge of our time—might offer . . . Not that I was expecting all the answers, much less definitive ones. But how could one ignore what humankind had managed to comprehend, over the past two centuries, about the detailed structure of things?

ꝏ

I found the study of classical physics mildly entertaining. Its concision gave it elegance. It was more coherent and definitely made more sense than the little formulas I'd had to rote-learn at school. Studying Einstein's discoveries on space and time filled me with amazement and exhilaration. My heart was beating faster.

But it was during my first encounter with the quanta that colored lights started firing up in my brain. I felt that I was getting to touch the incandescent matter of reality. A place where our assumptions and prejudices about reality are thrown into question.

My encounter with quantum theory began at the deep end: one-on-one with Paul Dirac's seminal book. It happened like this: at Bologna, I had taken Professor Fano's course on "Mathematical Methods for Physics." "Methods," that is, for us. The course required a topic to be developed individually in some depth and then presented to the class. I had chosen a small area that is now studied by everyone graduating in physics, but which at the time was not part of the curriculum: *group theory*. I went to speak to Professor Fano to ask him about what my presentation should contain. His answer was: "The fundamentals of group theory *and its application to quantum theory*." I sheepishly mentioned the fact that I had not yet taken any course on "quantum theory" and knew literally nothing about it. "Oh, really?" he said. "Then you'd better get studying quick."

He was joking.

But I didn't realize that he was joking.

I bought Dirac's book, in the gray Boringhieri edition. It smelled good. (I always sniff books before buying them: the smell of a book is decisive.) I shut myself up at home and studied it for a month. I also bought four other books and studied those as well.[36]

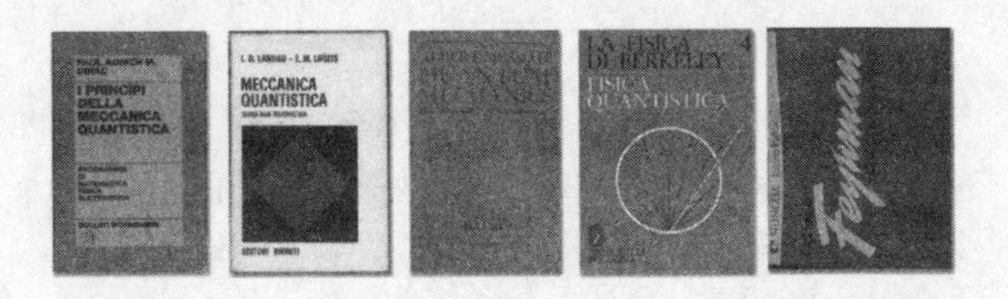

One of the most beautiful months I have ever experienced.

And the source of questions that have stayed with me for a lifetime. Questions which, after many years, much reading, countless discussions and doubts and uncertainties have led to the writing of *this* book.

In this chapter I delve into the strangeness of the quantum world. I describe a concrete phenomenon which captures that strangeness: a phenomenon that I have had the chance to observe in person. It is subtle, but it illustrates the crucial point. Then I list some of the ideas most discussed today that try to make sense of this strangeness.

I leave to the chapter after this one the idea I find most convincing—the *relational* interpretation of quantum mechanics. Readers impatient to get to it might want to leap there directly, skipping the perhaps entertaining but tangled circumlocutions of this one.

So what is it exactly that is so bizarre about quantum phenomena? The fact that electrons stay within certain orbits, and jump between them, is surely not the end of the world . . .

The phenomenon from which the strangeness of quanta derives is called "quantum superposition." A quantum superposition is when two contradictory properties are, in a certain sense, present together. An object could be here but at the same time elsewhere. It is what Heisenberg means by saying that "the electron no longer has a trajectory": the electron is found not to be only in one place or another. In a sense it is in both places. In the jargon, we say that an object can be in a "superposition" of positions. Dirac called this bizarre behavior the "principle of superposition." For him it was the theory's conceptual foundation.

But we need to be careful here: we never *see* a quantum superposition. What we see are consequences of the superposition. These consequences are called "quantum interference." It is the *interference* that we see, not the *superposition*.

Let's see how this works.

Quite a long time after I had studied it in books, I witnessed quantum interference with my own eyes. It was in the Innsbruck laboratory of Anton Zeilinger, a very kind Austrian with a big beard and the demeanor of a gentle bear—and one of the great experimental physicists who works wonders with quanta. He has been a pioneer of quantum computing, quantum cryptography and quantum teleportation. I'll describe what I saw in his lab, even if subtle, because it encapsulates why physicists are confused.

Anton showed me a table with a number of optical

instruments set up on it: a small laser, lenses, prism mirrors which separate the laser beam and then integrate it again, photon detectors and so on. A weak laser beam made up of a small number of photons is split into two parts, creating two separate paths—one, let's say, on the "right," and the other one on the "left." The two paths are reunited before becoming separated again and ending up in two detectors: one, let's say, "up" and the other "down."

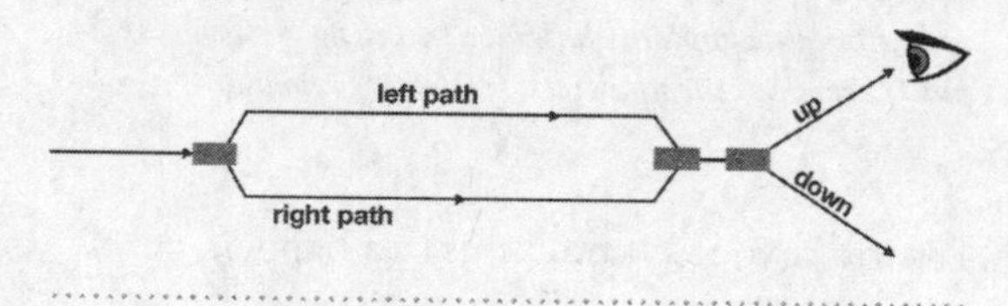

A beam of photons is separated into two by a prism, reunited and then divided again.

What I saw is this: If I blocked either one of the two paths (left or right) with my hand, half of the photons ended up in the down detector and half ended up in the up one (the two drawings on the left on the next page). But if I left both paths open, free of any impediment, *all of the photons ended up in the lower detector*: none in the one above it (the drawing on the right on the next page).

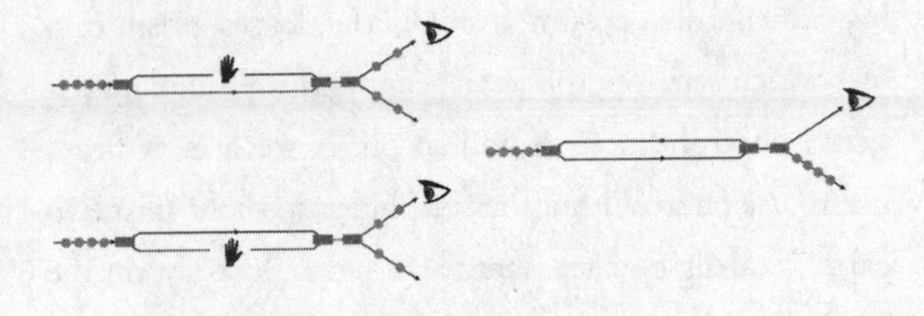

Quantum interference. If you block one of the two paths with your hand, half of the photons arrive in the up detector (left-hand image). If the two paths are unobstructed, all the photons end up in the down one (right-hand image). How does placing my hand in one of the paths cause photons that were traveling on the other path to move to the up detector? Nobody knows.

Try to ask yourself how this can happen.

There is something very peculiar going on. If half of the photons arrive at the up detector when one path is free, it would seem reasonable to expect that half of the photons should also arrive above when both paths are free. But this is not the case. In fact, *none* do.

How, by blocking one path, can my hand cause the photons traveling *on the other path* to go to the upper detector?

The disappearance of the photons from the upper detector when *both* paths are open is an example of *quantum interference*. There is *interference* between the two paths—the one on the left and the one on the right. When both are open, something happens that does not occur when photons pass on just one or the other of these paths: the photons going to the upper detector vanish.

According to Schrödinger's theory, the ψ wave of each photon separates into two parts: two wavelets. One wavelet follows the path on the right, one takes to the left. When they meet again, the ψ wave recomposes and takes the lower path. If I block one of the paths, the ψ wave does not recompose, and therefore behaves in a different way. That waves behave in this way is not strange: interference between waves is a well-known phenomenon. Light waves and ocean waves do the same things.

But we never see the wave; we always see individual photons *each passing on only one side*: either to the right or to the left. If we place photon detectors along the paths, these detectors never reveal "half photon." Each photon passes (entirely) on the left, or (entirely) on the right. Each photon behaves as if it passed through both trajectories, as waves do (otherwise there would not be interference), but if we look to see where the photon is, we always see it on just one path.

This is the *quantum superposition*: the photon passes

"both right and left." It is in a quantum superposition of two configurations: the one on the right and the one on the left. The consequence is the fact that the photon no longer goes upward, as it would if it passed along just one or the other of the two paths.

This is not all. There is something else, and it is truly challenging. If I *measure* which of the two paths the photon takes, the interference disappears!

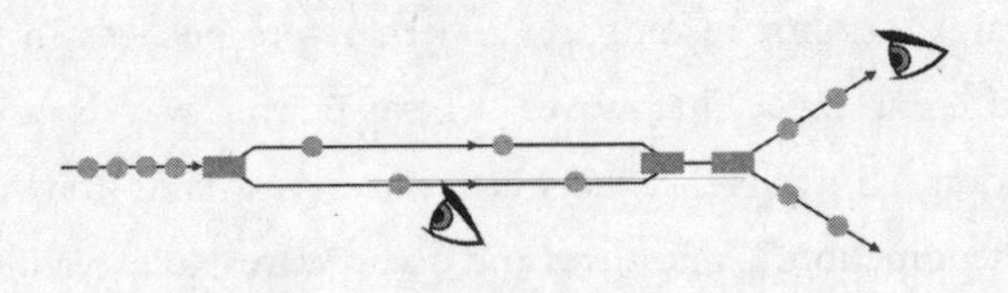

The very act of measuring which path the photon takes causes the interference to disappear! If we measure where they pass, half of the photons go upward again.

It seems that you need only to *observe* what is happening for it to change! Note the absurdity: if I *don't* look for where the photon passes, it always finishes below. But if I look at where it passes, it can end up above. The astonishing thing is that a photon can end up above *even if I haven't seen it*. That is to say, the photon changes trajectory due to the fact that I was waiting for it at the gate, on the side where it hasn't passed. Even if I haven't actually seen it!

What you read in textbooks on quantum mechanics is that if you *observe* where a photon passes, its ψ wave jumps entirely onto a path. Say you observe if the photon passes on the right path: if you see it, the ψ wave jumps completely to the right; but also if you *do not see it*, the ψ wave jumps! It jumps to the left. In both cases, there is no longer interference. The wave function "collapses," that is to say it leaps, converging in one point, the moment we observe it.

This is quantum superposition: the photon is, so to say, "on both paths." But if you search for it, it is only on one path.

Hard to believe.

And yet it happens: I saw it with my own eyes. Despite having studied it so much at university, seeing it and having, literally, a hands-on experience of it left me confused. Try yourself to think of a sensible explanation of this behavior. For a century now, we've all been trying.

If you find all this confusing, if you cannot make

head or tail of it, you are not alone. It is why Richard Feynman wrote that *nobody* understands quanta. (If instead what I have described seems perfectly clear, then it means that I have not been clear enough about it. For as Niels Bohr once said, you should "never express yourself more clearly than you are able to think."[37])

ꝉꝉ

Schrödinger illustrated this same enigma with a famous thought experiment: instead of a photon that takes a path on the right and a path on the left at the same time, he imagined a cat that is asleep and awake at the same time.[38]

It goes like this: a cat is shut in a box with a device where a quantum phenomenon has a one-in-two probability of happening. If it happens, the device opens a bottle of sleeping draught and the cat falls asleep.* The theory tells us that the ψ wave of the cat is in a "quantum superposition" of "cat-awake" and "cat-asleep," and it remains so until we actually see the cat.

*In the original version, the bottle contains not a sleeping gas but poison, and the cat did not fall asleep: it died. I prefer not to play around with the death of a cat.

So if it is truly described by its ψ, the cat is in a "quantum-superposition" of "cat-awake" and "cat-asleep."

This is different from saying that *we do not know* if the cat is awake or asleep, for the following reason: there are interference effects between cat-awake and cat-asleep (analogous to the interference effects between the two paths of Zeilinger's photons) that would not occur if the cat was either awake or asleep. They happen when the cat is in this quantum superposition of cat-awake and cat-asleep. As in the interference in Zeilinger's experiment, which occurs only if the photons "pass along both paths."

For a physical system as large as a cat, the interference is too difficult to observe.[39] But there is no convincing reason to doubt its reality. The cat is neither awake nor asleep. It is in this *quantum superposition* between cat-awake and cat-asleep.

But what does this mean?

How does a cat feel, being in a quantum superposition of cat-awake and cat-asleep? If you were in a quantum superposition between yourself-awake and yourself-asleep, then how, dear reader, would *you* feel? This is the riddle of quanta.

TAKING Ψ SERIOUSLY: MANY WORLDS, HIDDEN VARIABLES AND PHYSICAL COLLAPSES

To provoke a heated discussion at a physics conference dinner, you need only turn to the person next to you and casually ask: "So, in your opinion, is Schrödinger's cat awake *and* asleep?"

Discussions on the mysteries of quanta were lively in the 1930s, immediately after the birth of the theory. A famous debate on this topic between Einstein and Bohr went on for years through personal encounters, conferences, written works, letters . . . Einstein was resistant to the idea of relinquishing a more realistic image of phenomena. Bohr defended the conceptual novelty of the theory.[40]

In the 1950s, the problem was mostly ignored: the power of the theory was so spectacular that physicists tried to apply it in every possible field, without asking too many questions. But if you don't ask questions, you learn nothing.

By the 1960s, interest in the conceptual problems was on the rise again, curiously added to by the fascination within hippie culture for the alternative otherness of quanta.[41]

Today, discussions of quantum theory are frequent in departments of philosophy as well as in physics departments, and there are discordant opinions and perspectives. Some ideas are abandoned, others persist. The ideas that have withstood criticism give us ways of comprehending quanta, but each one of these ways has a high conceptual cost: each forces us to accept something outlandish. The judgment on the final balance of costs and benefits entailed by the various opinions on the theory is still open.

I expect that we will end up agreeing, as has happened with the other great scientific disputes that seemed irresolvable at the time. Is the Earth stationary, or does it move? (It moves.) Is heat a fluid, or is it the rapid movement of molecules? (It is the movement of molecules.) Do atoms really exist? (Yes.) Does the world consist only of "energy"? (No.) Do we have ancestors in common with apes? (Yes.) And so on . . . In this book, which is a chapter in the ongoing dialogue, I describe where it seems to me the debate is now, and in which direction we are going.

Before reaching, in the next chapter, the ideas that I find most convincing, namely the relational perspective, I summarize below the most discussed alternatives. They have come to be called "interpretations of quantum mechanics." In one way or another, they all require the acceptance of extreme possibilities: multiple universes, invisible variables, phenomena that have never been observed, and other such strange beasts. This is nobody's fault: it is the fundamental strangeness of the theory itself that forces us to resort to extreme solutions. The rest of this chapter is full of speculations. You may jump to the next chapter, where I get to what in my opinion is the heart of the matter; but if you want a panoramic view of the current discussion and strange ideas in play, here they are, and they are fun.

Many Worlds

The "Many Worlds" interpretation is currently fashionable in certain philosophical circles, as well as with some theoretical physicists and cosmologists. The idea is to take Schrödinger's theory seriously. In other words, *not* to interpret the ψ wave as a probability—but to see it instead as a real entity, effectively describing the world as it is.

In a certain sense, the idea is to discount the Nobel Prize awarded to Max Born, given to him for having understood that the ψ wave is *only* an evaluation of probability.

Schrödinger's cat, if this is how things stand, is actually described by its completely real ψ wave. Hence it is *actually* in a superposition of cat-awake and cat-asleep: both exist concretely. Why is it, then, that if I open the box, I see the cat either asleep or awake, and not both things at once?

Hold on tight. The reason for this, according to the Many Worlds interpretation, is that I, too, Carlo, am described by my ψ wave. When I observe the cat, my very real ψ wave interacts with the cat's, and my very real ψ wave also separates into two parts: one representing a version of myself that sees the cat awake, and one representing a version of myself that sees the cat asleep. Both, according to this perspective, are real.

Hence the total ψ now has two components: two

"worlds." The world has branched into two: one world in which the cat is awake, and I see the cat awake, and another world in which the cat sleeps, and I see him sleeping. There are now two versions of myself: one for each world.

So why is it that *I* see, for example, only the cat awake? The answer is that "I" am now just *one of two* versions of myself. In a parallel world that is equally real, equally concrete, there is a parallel me seeing the cat asleep. This is why the cat can be awake and at the same time asleep, but if I look at it, I see one thing only. Because if I look, I also become double.

Given that the ψ of Carlo interacts continuously with innumerable systems other than the cat, it follows that there is an infinity of other parallel worlds, equally existent, equally real, where an infinite number of copies of myself exist and experience every sort of alternative reality. This is the Many Worlds theory.

Does this sound crazy? It sounds crazy because it is.

Yet there are eminent physicists and eminent philosophers who maintain that this is the best possible reading of quantum theory.[42] They are not the crazy ones: the craziness lies with this incredible theory that has worked so well for a hundred years.

But in order to emerge from the fog around quantum theory, is it really worth giving credence to the real and concrete existence of infinite copies of ourselves, unknown and unobservable to us, hidden behind a gigantic universal ψ?

I have another problem with this interpretation of quantum theory. The gigantic, universal ψ wave that contains all the possible worlds is like Hegel's dark night in which all cows are black: it does not account, per se, for the phenomenological reality that we actually observe.[43] In order to describe the phenomena that we observe, other mathematical elements are needed besides ψ: the individual variables, like X and P, that we use to describe the world. The Many Worlds interpretation does not explain them clearly.

Hidden Variables

There is a way of avoiding an infinite multiplication of worlds and of copies of ourselves. It is provided by a group of theories called "Hidden Variables." The best of these was conceived of by de Broglie, the originator of the idea of waves of matter, and refined by David Bohm.

David Bohm is a scientist who had a difficult life because he was a communist on the wrong side of the Iron Curtain. Investigated during the era of McCarthyism in the United States, he was arrested in 1949 and briefly imprisoned. He was released—but sacked by Princeton anyway, which, hypocritically, feared for its reputation. He emigrated to South America. Fearing that he would end up defecting to the Soviet Union, the American embassy canceled his passport . . .

The idea behind Bohm's theory is simple: the ψ wave of an electron is a real entity, as in the Many Worlds interpretation; but in addition there is *also* the actual electron: a real material particle that has a definite position. There is only a single position, as in classical mechanics: no quantum superposition. The ψ wave evolves always following Schrödinger's equation, while the real electron moves in physical space, guided by the ψ wave.

Bohm devised an equation that showed how the ψ wave could effectively guide the electron.[44]

The idea is brilliant: the phenomena of interference are determined by the ψ wave that guides objects; but those objects themselves are not in a quantum superposition. They are always in a single position. The cat is awake, or it is asleep. But its ψ has both components: one corresponds to the "real" cat, the other is an "empty" wave without a real cat—but the empty wave can bring about an interference. Interfering, that is, with the wave of the real cat.

This is why we see a cat awake or asleep, and yet there are interference effects: the cat is in one state, but in the other state there is a part of its wave that generates interference.

This provides a good explanation of the Zeilinger experiment described above. Why, when I block *one* of the two paths, does my hand influence the movement of the photons passing along the *other* path? Because the electron passes along one path only, but its wave passes along both. My hand alters the wave that then guides the electron in a way that is different to how it would behave if my hand had not intervened. In this way, my hand alters the future behavior of the electron, even if the electron passes at a distance from my hand. It's a very good explanation.

The Hidden Variables interpretation brings quantum physics back into the same logical realm as classical physics: everything is deterministic and predictable. If we knew the position of the electron and the value of the wave, we could predict everything.

But it's not really as simple as this. As it happens, we *cannot ever* know the wave, because we never see it: we only see the electron.[45] Hence the behavior of the electron is determined by variables (the wave) that for us remain hidden. The variables are hidden in principle: we can *never* determine them. This is how the theory gets the name *Hidden* Variables.[46]

The price to be paid for taking this theory seriously is to accept the idea that an entire physical reality exists that is in principle inaccessible to us. Its sole purpose, when it comes down to it, is merely to comfort us with regard to what the theory does not tell us. Is it worth assuming the existence of an unobservable world, with no effect not already foreseen by quantum theory, only to assuage our fear of indeterminacy?

There are other difficulties as well. Bohm's interpretation is favored by some philosophers because it offers a conceptually clear framework. But it is liked less by physicists, because as soon as you try to apply it to something more complicated than a single particle, problems accumulate. The ψ wave of more particles, for example, is not the sum of the single particles: it is a wave that does not move in physical space, but in an abstract mathematical space.[47] The intuitive and clear image of reality that Bohm's theory provides in the case of a single particle is lost.

Even more serious problems occur when relativity is taken into account. The hidden variables of the theory violate relativity brutally: they determine a privileged (unobservable) reference system. The price of thinking that the world is made of variables that are always determined, as in classical physics, is not just accepting that these variables are forever hidden, but also that they contradict everything that we have learned about the world, precisely through classical physics. Can it really be worth such a cost?

Physical Collapse

There is a third way of considering the ψ wave to be real that avoids both Many Worlds and Hidden Variables: by thinking of the predictions of quantum mechanics as *approximations* that overlook something capable of rendering everything more coherent.

There could be a real physical process, independent of our observations, that happens *spontaneously* every so often and prevents the wave from scattering. This hypothetical mechanism, never directly observed to date, is called the "physical collapse" of the wave function. The "collapse of the wave function" would not happen because we observe it; it happens spontaneously. The more macroscopic the objects in question, the more rapidly it would occur.

In the case of the cat, the ψ would quickly leap by itself to one of the two configurations, and the cat would rapidly be asleep or awake. If so, regular quantum mechanics no longer applies for macroscopic entities such as cats.[48] Hence this type of theory gives predictions that deviate from those of usual quantum theory.

Various laboratories around the world have tried and are continuing to try to check these predictions in order to see who's right. For now, it is quantum theory that has always turned out to be right. Many physicists, including your humble author, would bet on quantum theory continuing to be right for a while yet.

ACCEPTING INDETERMINACY

The interpretations of quantum theory discussed so far avoid indeterminacy by taking ψ to be a real entity.[49] The cost is to add to reality things such as multiple worlds, inaccessible variables, or never-observed processes. But there is no reason to take the ψ wave so seriously.

ψ is not a real entity: it is a calculation tool. It is like weather forecasts, the profit predictions of a company, like horseracing odds.[50] Real events in the world happen in a probabilistic way, and the quantity ψ is our way of calculating the probability of them occurring.

Interpretations of the theory that do not take the ψ waves so seriously are called "epistemic," because they interpret ψ only as a summary of our knowledge (ἐπιστήμη) of what happens.

An example of this way of thinking is *QBism*. QBism takes quantum theory as it finds it, without seeking to "complete" the world: without hypothesizing other worlds, hidden variables or processes for which we have no evidence.

The idea is that ψ is only the information that we ourselves have about the world. It describes "that which *we* know about the world." The information I have grows when I make an observation. That is why ψ changes when we observe: not because something happens in the external world, but just because the information that we have changes. Our forecast of the weather changes if we look at a barometer: not because the weather promptly changes the moment that we consult the barometer, but because we have suddenly learned something that we did not know before.

QBism gets its name from "Quantum-Bayesianism."

(Thomas Bayes was an eighteenth-century Presbyterian minister who studied probability.) But the word "QBism" alludes also to the *Cubism* of artists such as Georges Braque and Pablo Picasso—the influential style of painting that was shaped in Europe in the same years that quantum theory was developing. Cubism and quantum theory both moved away from the idea that the world is representable from a single perspective. In the first decades of the twentieth century, it is the whole of European culture that no longer thinks we can represent the world in a simple and complete way. The anthropologist Claude Lévi-Strauss understands that to study a culture is to change it; Sigmund Freud understands that doctors cannot evaluate patients' minds without affecting them. In Italy, between the years 1909 and 1925, the years during which quantum theory is born, Luigi Pirandello writes *One, No One and One Hundred Thousand* (1926), which speaks of the splintering of reality into the points of view of countless observers . . .

QBism abandons a realistic image of the world, beyond what we can see or measure. The theory gives us the probability that we will see something, and this is all that it is legitimate to say. It is not legitimate to

say anything about the cat or the photon when we are not actually observing them.

QBism holds a drastically instrumental conception of science: the theory gives predictions only about what a subject can see. I think that science is not just about making predictions. It also provides us with a vision of reality, a conceptual framework for thinking about things. It is this aspiration that has made scientific thought so effective. If the objective of science was solely to make predictions, Copernicus would not have discovered anything new with respect to Ptolemy. His astronomical predictions are no better than Ptolemy's. But Copernicus found a key with which to rethink everything, to reach a new level of understanding.

The weakness of QBism, in my opinion—and this is the turning point in this whole discussion—is that QBism anchors reality to a subject of knowledge, an "I" that knows, as if it stood outside nature. Instead of seeing the observer as a part of the world, QBism sees the world reflected in the observer. In so doing, it leaves behind naive materialism but ends up falling into an implicit form of idealism.[51] The crucial point that QBism disregards, I believe, is that *the observer himself can be observed*. We have no reason to doubt that every real observer is himself described by quantum theory.

If I observe an observer, I see things that the observer does not. I deduce, by reasonable analogy, that therefore there are also things that I, as an observer, do not see. I want a theory of physics that accounts for the structure of the universe, that clarifies what it is to be an observer in the universe, not a theory that makes the universe depend on me observing it.

In the end, all of the interpretations of quantum theory outlined in this chapter repeat the debate between Schrödinger and Heisenberg: between a "wave mechanics" that tries at all costs to avoid indeterminacy and probability in the world, and the radical leap of the "boys' physics" that seems to depend too much upon the existence of a subject who "observes." This chapter has introduced us to an array of intriguing ideas, but it has not allowed us to make a real step forward.

Who is the subject who knows and retains information? What is the information that the subject has? What is, after all, this subject who observes? Does it escape from the laws of nature, or is it subjected to and described by natural laws? If it is part of nature, why treat it as in any way special?

This question, one of countless reformulations of the central question raised by Heisenberg—what is an observation? what is an observer?—finally brings us to the main concept of this book: *relations*.

III

IS IT POSSIBLE THAT SOMETHING IS REAL IN RELATION TO YOU BUT NOT IN RELATION TO ME?

Where I finally talk about relations.

THERE WAS A TIME WHEN THE WORLD SEEMED SIMPLE

At the time Dante was writing, in Europe we thought of the world as the blurred mirror of a great celestial hierarchy: a great God and His spheres of angels carried the planets in their course across the heavens and participated with trepidation and love in the lives of a fragile humanity which oscillated, at the center of the universe, between adoration, rebellion and guilt.

Then our views changed. In the centuries that followed, we understood aspects of reality, discovered hidden grammars, found strategies for our objectives. Scientific thinking has woven a complex edifice of knowledge. Physics has played a determining and unifying role, providing a clear image of reality: a vast space where particles run, pushed and pulled by forces. Faraday and Maxwell added the electromagnetic "field": an entity diffused in space through which distant bodies exercise influence upon each other. Einstein completed the picture by showing that gravity is also carried by a "field": a field that is the very geometry of space and time. The synthesis is tremendously lucid and beautiful.

Reality is a luxuriant stratification: snow-covered mountains and forests, the smiles of friends, the rumble of the subway on dirty winter mornings, our insatiable thirst, the dance of our fingers across a laptop keyboard, the taste of bread, the sorrow of the world, the night sky, the immensity of the stars, Venus shining alone in the ultramarine blue of twilight . . . Behind this disordered veil of appearances, we thought we had found the deep weave, the hidden order. This was the time in which things seemed simple.

But the great hopes of the insignificant mortal critters that we are too often turn out to be short-lived dreams. The conceptual clarity of classical physics has been swept away by quanta. Reality is decidedly *not* how it is described by classical physics.

This was an abrupt awakening from the pleasant sleep in which we had been cradled by the illusions of Newton's success. But it was a reawakening that connects us back to the beating heart of scientific thinking, which is not made up of acquired certainties: it is thinking constantly in motion, the power of which is precisely the capacity to always question everything and begin over again, to be fearless in subverting the order of the world in the search for a more efficient one, only to then put a further question mark over everything, to subvert it all over again.

Not to fear rethinking the world is the power of science: ever since Anaximander removed the foundations on which the Earth rested, Copernicus launched it to rotate in the sky, Einstein dissolved the rigidity of space and of time, and Darwin demolished the separateness of humanity . . . reality is constantly being redrawn in images that are increasingly effective. Step by step, the fabulous strangeness and beauty of reality is unveiled. The courage to radically reinvent the world: this was the subtle fascination of science that first captivated me as a rebellious adolescent . . .

RELATIONS

In a physics laboratory, where we study a small object such as an atom or a photon of Zeilinger's lasers, it is clear who the *observer* is: it is the scientist who prepares, observes and measures the quantum object, who deploys their instruments of measurement, detecting the light emitted from the atom or the place where the photons arrive.

But the vast world is not made up of scientists in laboratories, or instruments of measurement. What is an observation, when there is no scientist observing? What does quantum theory tell us, where there is no one measuring? What does quantum theory tell us about what happens in another galaxy?

The key to the answer, I believe, and the keystone of the ideas in this book, is the simple observation that scientists, and their measuring instruments as well, are all part of nature. What quantum theory describes, then, is *the way in which one part of nature manifests itself to any other single part of nature.*

At the heart of the "relational" interpretation of quantum theory is the idea that the theory does not describe the way in which quantum objects manifest themselves *to us* (or to special entities that do something special denoted "observing"). It describes how every physical object manifests itself to any other physical object. How any physical entity acts on *any* other physical entity.

We think of the world in terms of objects, things, entities (in physics, we call them "physical systems"): a photon, a cat, a stone, a clock, a tree, a boy, a village, a rainbow, a planet, a cluster of galaxies . . . These do not exist in splendid isolation. On the contrary, they do nothing but continuously act upon each other. To understand nature, we must focus on these interactions rather than on isolated objects. A cat listens to the ticking of a clock; a boy throws a stone; the stone moves the air through which it flies, hits another stone and moves that, presses into the ground where it lands; a tree absorbs energy from the sun's rays, produces the oxygen that the villagers breathe while watching the stars, and the stars run through the galaxies, pulled by the gravity of other stars . . . The world that we observe is continuously interacting. It is a dense web of *interactions*.

Individual objects *are* the way in which they interact. If there was an object that had no interactions, no effect upon anything, emitted no light, attracted nothing and repelled nothing, was not touched and had no smell . . . it would be as good as nonexistent. To speak of objects that never interact is to speak of something—even if it existed—that could not concern us. It is not even clear what it would mean to say that such objects "exist." The world that we know, that relates to us, that interests us, what we call "reality," is the vast web of interacting entities, of which we are a part, that manifest themselves by interacting with each other. It is with this web that we are dealing.

One of these entities is a photon observed by Zeilinger. But another is Anton Zeilinger himself. Zeilinger is an entity—like a photon, a cat or a star. You, reading these lines, are another such entity, and I—as I write them on a Canadian winter morning with the sky through the window of my study still dark, and an amber-colored kitten purring nestled between myself and the computer on which I'm working—I am also an entity like the others.

If quantum theory describes how a photon manifests itself to Zeilinger, and these are two physical systems, it must also describe the way in which *any* object manifests itself to *any other* object.

There are particular systems that are "observers" in a strict sense of the term: have sense organs and memory, work in a laboratory, interact with a large environment, are macroscopic. But quantum mechanics does not describe only these: it describes the elementary and universal grammar of physical reality underlying not just laboratory observations but every type and instance of interaction.

If we look at things in this way, there is nothing special in the "observations" introduced by Heisenberg: *any* interaction between two physical objects can be seen as an observation. We must be able to treat any object as an "observer" when we consider the manifestation of other objects to it. Quantum theory describes the manifestations of objects to one another.

The discovery of quantum theory, I believe, is the discovery that the properties of any entity are nothing other than the way in which that entity influences others. It exists only through its interactions. Quantum theory is the theory of how things influence each other. And this is the best description of nature that we have.[52]

It is a simple idea, but it has radical consequences that open the conceptual space required to understand quanta.

No Interaction, No Properties

Bohr speaks of the "impossibility of neatly separating the behavior of atomic systems from their interaction with the measuring device used to define the conditions under which the phenomenon appears."[53]

When he wrote this, in the 1940s, the applications of the theory were confined to the laboratories that measured atomic systems. Almost a century later, we know that the theory is valid for every object in the universe. We need to amend "atomic systems" to "all objects," and "interaction with measuring equipment" to "interaction with any other thing whatsoever."

Revised in this way, Bohr's observation captures the discovery that forms the basis of the theory: the impossibility of separating the properties of an object from the interactions in which these properties manifest themselves and the objects to which they are manifested. The properties of an object *are* the way in which it acts upon other objects; reality is this web of interactions. Instead of seeing the physical world as a collection of objects with definite properties, quantum theory invites us to see the physical world as a net of relations. Objects are its nodes.

The first radical consequence is that to attribute properties to something when it does *not* interact is superfluous and may be misleading. It is talking about something that has no meaning, for *there are no properties outside of interactions.*[54]

This is the significance of Heisenberg's original intuition: to ask what the orbit of an electron is when it is not interacting with anything is an empty question. The electron does not follow an orbit because its physical properties are only those that determine how it affects something else, for instance, the light that it emits *when* it is interacting. If the electron is not interacting, there are no properties.

This is a radical leap. It is equivalent to saying that everything consists *solely* of the way in which it affects something else. When the electron does not interact with anything, it has no physical properties. It has no position; it has no velocity.

Facts Are Relative

The second consequence is even more radical.

Suppose that you are the cat in Schrödinger's thought experiment. You are shut in a box and a quantum mechanism has a one in two probability of releasing the sleeping drug. *You* perceive whether the drug has been released or not released. In the first case, you sleep; in the second, you remain awake. *For you* the drug was delivered or it was not delivered. There are no doubts. *As far as you are concerned*, you are asleep or you are awake. You are certainly not both at once.

I, on the other hand, am outside the box and do not interact either with the bottle of sleeping draught or with you. Later on, *I* can observe interference phenomena between you-awake and you-asleep: phenomena that would not have been produced if I had seen you asleep, or if I had seen you awake. In this sense, *for me* you are neither asleep nor awake. This is what it means to say that you are "in a superposition of sleeping and waking."

For you the soporific is released or not, and you are asleep or awake. *For me* you are neither awake nor asleep. For me, "there is a quantum superposition." For you, there is the reality of being awake, or of not being so. The relational perspective allows *both things to be true*: each relates to interactions with respect to distinct observers—you and me.

Is it possible that a fact might be real with respect to you and not real with respect to me?

Quantum theory, I believe, is the discovery that the answer is yes. *Facts that are real with respect to an object are not necessarily so with respect to another.** A property may be real with respect to a stone, and not real with respect to another stone.[55]

*The problem of quantum mechanics is the apparent contradiction between two laws of the theory: one describes what happens in a "measurement," and the other in the "unitary" evolution, namely when there is no measurement. The relational interpretation is the idea that both are correct: the first regards the events relative to the systems in interaction, the second regards the events relative to other systems.

THE RAREFIED AND SUBTLE WORLD OF QUANTA

I hope that I have not lost my reader in the last few subtle but essential paragraphs. The gist is that the properties of objects exist only in the moment of their interactions, and they can be real with respect to one object and not with respect to another.

The fact that some properties exist only with respect to something else should not overly surprise us. We already knew as much. Speed, for example, is a property that an object has *relative to another object.* If you walk along the deck of a ferry, you have a speed relative to the ferry, a different speed relative to the water in the river, a different one relative to the Earth, another relative to the Sun, another again relative to the galaxy—and so on, endlessly. Speed does not exist without being anchored (implicitly or explicitly) to something else. Speed is a notion regarding two objects (you and the ferry, you and the Earth, you and the Sun . . .). It is a property that exists only with respect to something else. It is a *relation* between two entities.

There are many similar examples: since the Earth is a sphere, "up" and "down" are not absolute notions but *relative* to where we find ourselves on the Earth. Einstein's special relativity is the discovery that the notion of simultaneity is relative, and so on. The discovery of quantum theory is only slightly more radical: it is the discovery that *all* the properties (variables) of *all* objects are relational, just as in the case of speed.

Physical variables do not describe things: they describe the way in which things manifest themselves to each other. There is no sense in attributing a value to them if it is not in the course of an interaction.

The ψ wave is the probabilistic calculation of where and how an event *relative to us* might occur.[56] The wave as well is therefore a perspectival quantity. An object does not have one ψ wave, it has one with respect to every other object with which it interacts. Events that take place in relation to one thing do not influence the probability of events that occur in relation to others.* The "quantum state" ψ is always a relative state.[57]

The world is the network of relative facts: relations realized when physical entities interact. A stone collides

*This is the central technical feature of the relational interpretation. The probability of events realized *with respect to a system* is determined by the transition amplitudes that are functions of events realized with respect to *the same system*, and not by the amplitudes that are functions of events realized *with respect to other systems*.

with another stone. The light from the sun reaches my skin. You read these lines.

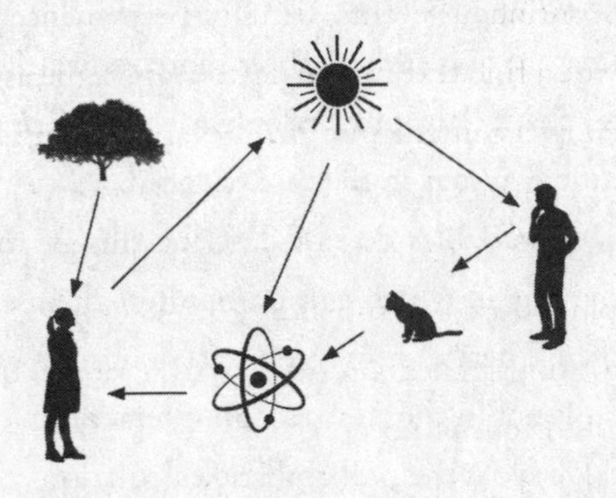

ꝉꝉ

The world that emerges from these considerations is a rarefied one. A world in which, rather than independent entities with definite properties, there are entities that have properties and characteristics only with regard to others, and only when they interact. A stone does not have a position in itself: it only has a position in relation to another stone with which it collides. The sky does not in itself have any color: it has color with respect to my eyes when they look at it. A star does not shine in the sky as an autonomous entity: it is a node in the network of interactions that forms the galaxy in which it resides.

The quantum world is more tenuous than the one imagined by the old physics; it is made up of happenings, discontinuous events, without permanence. It is a world with a fine texture, intricate and fragile as Venetian lace. Every interaction is an event, and it is these light and ephemeral events that weave reality, not the heavy objects charged with absolute properties that our philosophy posited in support of these events. "There are *fewer* things in heaven and earth, Horatio, Than are dreamt of in your philosophy . . ."

The life of an electron is not a line in space: it is a dotted manifestation of events, one here and another there. Events are punctiform, discontinuous, probabilistic, relative.

In *Cosmological Koans*, Anthony Aguirre describes this disconcerting conclusion in the following way:

> An electron is a particular type of *regularity* that appears among measurements and observations that *we make*. It is more pattern than a substance. It is *order* . . . Thus we arrive at a strange place. We break things down into smaller and smaller pieces, but then the pieces, when examined, are not there. Just the arrangements of them are. What then, are *things*, like the boat, or its sails, or

> your fingernails? What *are* they? If things are forms of forms of forms of forms, and if forms are order, and order is defined by us . . . they exist, it would appear, only as created by, and *in relation to*, us and the Universe. They are, the Buddha might say, emptiness.[58]

The solidity of the world to which we have become accustomed in our daily lives does not reflect the actual grain of reality: it is the result of our macroscopic vision. A lightbulb does not emit continuous light, it emits a hail of evanescent photons. At small scale, there is no continuity, or fixity, in the real world: there are discrete events, interactions, gapped and discrete.

Schrödinger had fought tooth and nail against quantum discontinuity, against Bohr's quantum leaps, against Heisenberg's world of matrices: he wanted to defend the image of continuous reality provided by classical intuition. But in the end even he capitulated, decades after the clashes of the 1920s, and admitted defeat. Schrödinger's words after the ones mentioned earlier ("There was a moment when the creators of wave mechanics nurtured the illusion of having eliminated the discontinuities in quantum theory") are clear and definitive:

> It is better to consider a particle not as a permanent entity but rather as an instantaneous event. Sometimes these events form chains that give the illusion of being permanent, but only in particular circumstances and only for an extremely brief period of time in each individual case.[59]

tt

The Many Worlds and Hidden Variables interpretations sketched in the previous chapter seek to "fill" the world with additional realities beyond what we see, to recover the "plenitude" of the classical world, to exorcise the indeterminacy of quanta. The cost of these approaches is to postulate a world full of invisible things. The relational perspective takes the theory as it is—it is the best theory that we have—with its sketchy description of the world, and accepts indeterminacy,* as QBism does. But while QBism is about the information of a

*In the Many Worlds interpretation, every time I observe an event, there is "another me" who observes something different. In Bohm's theory, only one of the two components of ψ includes me: the other is empty. The relational interpretation disconnects what I observe from what another observer observes: if I am the cat, I am asleep or awake, but this does not prevent interference phenomena, because there is no element of reality actualized with respect to other observers that would limit such interference. The observation I have made is an event relative to me, not to others.

subject, the relational understanding of quantum theory is about the structure of the world.

To understand quantum theory, we need to modify the grammar of our understanding of reality, as when Anaximander understood that the true shape of the Earth changed the grammar of notions of what is "up" and "down."[60] Objects are described by variables that assume value when interacting, and this value is determined in relation to the objects in the interaction, not to others. An entity is *one, no one and one hundred thousand*.

The world fractures into a play of points of view that do not admit of a univocal, global vision. It is a world of perspectives, of manifestations, not of entities with definite properties or unique facts. Properties do not reside in objects, they are bridges between objects. Objects are such only with respect to other objects, they are nodes where bridges meet. The world is a perspectival game, a play of mirrors that exist only as reflections of and in each other.

This phantasmal world of quanta is our world.

IV

THE WEB OF RELATIONS THAT WEAVES REALITY

In which I speak about how things speak to one another.

ENTANGLEMENT

There is a subtle, beguiling quantum phenomenon that embodies the radical interdependence of things. The most enchanted and dreamy of the quantum phenomena. *Entanglement.*

Entanglement is the strangest of all strange quantum phenomena, the one that takes us furthest away from our old understanding of the world. But it is also something general, which in a sense weaves the very structure of reality.

It is the phenomenon by which two distant objects maintain a kind of weird connection, as if they continued to speak to each other from afar. They remain, as we say, "entangled," linked together. Like two lovers who can guess each other's thoughts when apart. It has been well verified in laboratories. Chinese scientists, led by Juan Yin, have succeeded in producing two entangled photons on a satellite called Micius and sending them, still entangled, to two stations at a distance of thousands of kilometers from each other on Earth.[61]

Let's see how this works.

First, two entangled photons have *correlated* features: if one is red, the other will be red; if one is blue, the other one will be blue as well. Nothing strange so far. If I separate a pair of gloves and send one to Vienna and one to Beijing, the one that arrives in Vienna will be of the same color as the one that ends up in Beijing: they are correlated.

The strangeness emerges when a pair of photons sent to Vienna and Beijing, respectively, are in a quantum superposition. For instance, they could be in a superposition of a configuration in which both are red, and one in which both are blue. Each photon may reveal itself as either red or blue the moment it is observed, but if one is found to be blue, then the other—far away—will also be blue.

The puzzling aspect is this: How can they turn out to be the *same* color? The theory states that each of the two photons is neither definitively red nor definitively blue until it interacts. It states that the color is determined randomly when we look. But if this is the case, how can the color randomly determined in Beijing be the same as the one randomly determined in Vienna? If I toss a coin in Beijing and another in Vienna, the results are independent of one another, they are not correlated: there is nothing causing it to be heads in Vienna every time it is heads in Beijing.

There seem to be only two possible explanations. The first is that a signal of the color of one photon travels extremely rapidly to the other, far-off photon; when a photon decides to be either red or blue, it communicates this instantly in some way to its distant brother photon. The second, more reasonable possibility is that the color was already determined at the moment of the separation, just as in the case of the gloves, even if we were not aware of it. (Einstein expected this to be the case.)

Neither of these explanations works. The first implies an impossibly rapid communication over too great a distance, against all we know about the structure of space-time, which prevents such rapid signals. In fact, there is no way of using entangled objects to send signals. Hence the correlation is not related to a rapid signal transmission.

As for the other possibility—that the photons, like the gloves, already "knew" before being separated that they would both be blue or both red—it has been excluded as well. It was excluded by acute observations made in a brilliant article written in 1964 by a physicist from Belfast named John Bell.[62] Bell's job was in particle physics and particle accelerator design; understanding quantum theory was a matter of personal curiosity for him, at a time when almost no one cared about the issue. Yet today, he is celebrated for his influence on the foundation of quantum physics.

With reasoning that is elegant, subtle and very technical, Bell showed that if all the correlated properties of the two photons had been determined from the moment of separation (instead of being determined by chance at the moment of observation), precise consequences would follow (today called "Bell inequalities") that are contradicted by what we actually observe. The correlations are definitely *not* determined from the outset.[63]

It seems, then, like a puzzle without a solution. How can two entangled particles make the same decision without previous agreement and without sending each other messages? What is it that connects them?

ꝺꝺ

My good friend Lee recounts that as a young man he lay on his bed for hours on end looking at the ceiling, after he had studied entanglement. He was thinking about how each atom in his body must have interacted in some distant past with so many other atoms in the universe. Every atom in his body had to be entangled with billions of atoms dispersed throughout the galaxy . . . He felt a connectedness with the cosmos.

Entanglement shows that reality is definitely other than how we had conceived of it. Even if we know all that can be predicted about one object and another object, we still cannot predict everything about the two objects together.[64] The relationship between two objects is not something contained in one or the other of them: it is something more besides.[65]

This interconnection between all the components of the universe is disconcerting.

ᵵᵵ

Let's return to the puzzle: How do two entangled particles behave in the same way without having made up their minds beforehand and without communicating at a distance?

The relational perspective offers a solution, but one that shows just how radical that perspective is.

The solution lies in remembering that properties exist only in relation to something else. The measurement of the color of the photon performed in Beijing determines its color with respect to Beijing. But *not with respect to Vienna*. And vice versa. Since there is no physical object that sees both colors at the moment in which the two measurements are made, *it makes no sense to ask whether the two results are the same or not*. It is meaningless, because there is no object with respect to which the sameness is realized.

Only God can see the two places at the same moment—but God, if She exists, does not tell us what She sees. What She sees is irrelevant to reality. We cannot rely upon the existence of something that only God can see. We cannot assume that both colors exist, because there is nothing *with respect to which* both can be determined. Only the properties that exist in relation to something are real: the *combination* of the two colors does not exist in relation to anything.

We *can* compare the two measurements, in Beijing and Vienna, but the comparison requires an exchange of signals: the two laboratories can send each other emails or talk on the phone. An email takes time, as does a voice on the phone: nothing travels instantaneously, and an exchange of signals is an interaction, where new elements of reality come about.

When the result of the measurement in Beijing arrives in Vienna, by email or phone, *only then* does it become real with respect to Vienna as well. But at this point there is no longer a mysterious remote signal: with respect to Vienna, the concretization of the color of the Beijing photon occurs only when the signal containing the information arrives.

At the moment a measurement is made in Beijing, everything remains in quantum superposition *with respect to Vienna*. The equipment making the measurements, the scientists reading them, the notebooks in which they are written down, the messages in which the results of the measurements are conveyed, *are all quantum objects* themselves. Until they communicate with

Vienna, their condition *with respect to Vienna* is not determined: with respect to Vienna, they are all like the cat in a superposition of awake and asleep. They are in a quantum superposition of one configuration in which they have measured blue and one in which they have measured red.

The same is true in Vienna with respect to Beijing. For both, the correlations do not become real until signals arrive. In this way, we can understand the correlations without recourse to magically exchanged signals or predetermined results.

This is a solution to the puzzle, but it comes at a cost: no universal set of facts exists. There exist facts relative to Beijing, and facts relative to Vienna, and the two *do not match*. Facts relative to one observer are not facts relative to another. It is a shining example of the relativity of reality.

The joint properties of *two* objects exist only in relation to a *third*. To say that two objects are correlated means to articulate something with regard to a *third* object: the correlation manifests itself when the two correlated objects *both* interact with this third object, which can check.

The apparent incongruity raised by what seemed

like communication at a distance between two entangled objects was due to neglect of this fact: the existence of a third object that interacts with both the systems is necessary to give reality to the correlations. Everything that manifests itself does so *in relation to something.* A correlation between two objects is a property of the two objects—like all properties, it exists only in relation to a further, third object.

Entanglement is not a dance for two partners, it is a dance for three.

THE DANCE FOR THREE THAT WEAVES THE RELATIONS OF THE WORLD

Zeilinger looks at a photon and sees it is red. A thermometer measures the temperature of a cake being baked. A measurement is an interaction between one object (the photon, the cake) and another (Zeilinger, the thermometer). At the end of the interaction, one object has gathered information about another object. The thermometer has acquired information about the temperature of the cake. This means that there is a *correlation* between the

thermometer and the cake. After the measurement, if the cake is cold, then the thermometer indicates coldness (its column of mercury is low); if the cake is hot, the thermometer indicates this heat (the column of mercury is high). The temperature and the thermometer have become like the two photons: *correlated*.

Now, if the cake was in a quantum superposition of different temperatures, then with respect to the thermometer, the cake has manifested one of its properties (temperature) in the course of the interaction. But with respect to a third system of any kind that does not participate in this interaction, no property has manifested itself; the cake and the thermometer become entangled.

This is what happens with Schrödinger's cat. With respect to the cat, the sleeping draught appears, or it does not. With respect to me, having not yet opened the box, the bottle of sleeping draught and the cat are entangled: a quantum superposition of open-bottle/cat-asleep and closed-bottle/cat-awake.

Entanglement is therefore far from being a rare phenomenon that occurs only in particular situations: it is what happens, generically, in an interaction when this interaction is considered in relation to a system external to it.

From an external perspective, any manifestation of one object to another, which is to say any property, is a correlation; it is an entanglement between an object and another.

Entanglement, in sum, is none other than the external perspective on the very relations that weave reality: the manifestation of one object to another, in the course of an interaction, in which the properties of the objects become actual.

ꝥ

You look at a butterfly and see the color of its wings. In relation to me, a relation is established between you and the butterfly: the butterfly and you are now in an entangled state. Even if the butterfly moves away from you, the fact remains that if I look at the color of its wings and ask you which color you have seen, I will find that our answers match . . . even if it is not impossible that there will be subtle interference phenomena with the configuration whereby the butterfly is a different color.

All the information that we have about the world, considered externally, is in these correlations. Since all properties are relative properties, everything in the world does not exist other than in this web of entanglement.

But there is method in this madness. If I know that you have looked at the butterfly's wings, and you tell me that they were blue, I know that if I look at them I will see them as blue: this is what the theory predicts, *despite the fact that properties are relative.*[66] The fragmentation of points of view, the multiplicity of perspectives opened up by the fact that properties are only relative, is repaired, made coherent, by this consistency, which is an intrinsic part of the grammar of the theory.[67] This consistency is the basis of the intersubjectivity that grounds the objectivity of our communal vision of the world.

The wings of the butterfly will always be the same color for all of us.

INFORMATION

I close this second part of the book with a comment on the role of *information* in quantum theory. Words are never precise: the variegated cloud of meanings that they carry about with them is their expressive power. But it also generates confusion, "'cause you know sometimes words have two meanings." The word "information" that I used a few lines ago is a word packed with ambiguity. It is used to mean quite different things in different contexts.

It is often used to refer to something that has *meaning*. A letter from our father is "rich in information." In order to decipher this type of information, you need a mind that understands the *meaning* of the sentences in the letter. This is a "semantic" notion of information, that is to say, one that is linked to meaning.

But there is also a usage of the word "information" that is much simpler and has nothing to do with semantics or the mind: it comes directly from physics, where we do not speak of either minds or meanings. It is what I did above when writing that the thermometer "has information" on the temperature of the cake, in order to say only that when the cake is cold the thermometer indicates cold, and when the cake is hot the thermometer registers heat.

This is the simple and general sense of the word "information" used in physics. If I drop two coins, there are four possible results (heads-heads, heads-tails, tails-heads and tails-tails). But if I glue the two coins to a piece of transparent plastic, both face up, and let them drop, there are no longer four possible outcomes but just two: heads-heads and tails-tails. Heads for one coin implies heads for the other. In the language of physics, we say that the sides of the two coins are "correlated." Or that the sides of the two coins "have information about each other." If I see one, this "informs" me about the other.*

To say that a physical variable "has information" on another physical variable in this sense is simply to say that there is a tie of some kind (a common history, a physical link, the glue on the plastic sheet) due to which the value of one variable implies something about the value of the other.[68] This is the meaning of the word "information" that I am using here.

I hesitated to speak of information in this book, because the word is so ambiguous: everyone tends instinctively to read into it what they will, and this becomes an obstruction to understanding. But I have taken the risk of including it because the concept of information is important for quanta. Please remember that "information" is used here in a physical sense that has nothing to do with anything mental, or with semantics.

---

*Two variables have *relative information* if they can be in fewer states than the product of the number of states that each can be in.

This clarified, here is the point: it is possible to think of quantum physics as a theory of information (in the sense outlined) that systems have about one another. The properties of an object can be thought of, as we have seen, as the establishment of a correlation between two objects, or rather as *information* that one object has on another.

The same is true in classical physics. But this language allows us to pinpoint the difference between classical and quantum physics. This can be summarized in two general facts that radically differentiate quantum physics from classical physics, encapsulating the novelty of the quanta:[69]

1. *The maximal amount of relevant information about an object*[70] *is finite.*
2. *It is always possible to acquire new relevant information about any object.*

These two facts are so basic that they are called "postulates." At first glance the two postulates appear to contradict each other. If the information is finite, how is it always possible to obtain more of it? The contradiction is only apparent, however, because the postulates refer to "relevant" information. Relevant information is that which counts for predicting the future behavior of the object. When we acquire new information, part of the old information becomes "irrelevant": that is to say, it does not change what can be said about the future.[71]

Quantum theory is summed up in these two postulates.[72] Let's see how.

1. Information Is Finite: Heisenberg's Principle

If we could know with infinite precision all the physical variables that describe a thing, we would have infinite information. But this is impossible. The limit is determined by the Planck constant $\hbar$.[73] This is the meaning of Planck's constant. It is the limit up to which we can determine physical variables.

Heisenberg brought this crucial fact to light in 1927, shortly after having conceived the theory.[74] He showed that if the precision with which we have information on the position of something is ΔX, and the precision with which we have information on its speed (multiplied by its mass) is ΔP, the two precisions cannot both be arbitrarily good. They cannot both be too close to zero. Their product cannot be smaller than a minimum quantity: half the Planck constant. As a formula, it is this:

$$\Delta X \, \Delta P \geq \hbar/2$$

This reads: "Delta *X* times Delta *P* is always greater than or equal to h-bar divided by two." This general property of reality is called "Heisenberg's uncertainty principle." It applies to everything.

An immediate consequence is *granularity*. Light, for instance, is made of photons or grains of light, because portions of energy that were even more minute than this would violate this principle: the electric field and the magnetic field (that are like *X* and *P*, for light) would both be too determined and would violate the first postulate.

2. Information Is Inexhaustible: Noncommutativity

The uncertainty principle does not mean that we cannot measure the position of a particle with great precision and *then* its speed very precisely as well. We can. But after the second measurement, the position will no longer be the same: measuring the speed *loses information* on the position, so that if we measure it again, we will find it changed.

This follows from the second postulate, which says that even when we have gathered maximum information about an object, it is still possible to learn something unexpected about it. The future is not determined by the past: the world is *probabilistic.*

Since measuring P alters X, measuring X first and then P gives different results than measuring P and then X. Hence in mathematics "first X and then P" is necessarily different from "first P and then X."[75] This is precisely the property that characterizes *matrices*: order counts.[76] Remember the single new equation introduced by quantum theory?

$$XP - PX = i\hbar$$

This tells us precisely about the importance of order: "first X and then P" is different from "first P and then X." How different? By an amount that depends on

Planck's constant: the scale of quantum phenomena. This is why Heisenberg's matrices work: because they allow the order in which information is acquired to be taken into account.

Heisenberg's principle—that is, the equation on page 105—follows with a few steps from this last equation, which therefore summarizes everything. This equation translates into mathematical terms both the postulates of quantum theory. Vice versa, the two postulates express its physical significance.

In Dirac's version of quantum theory there is not even a need for matrices: everything may be obtained by simply using "noncommutative" variables, which is to say, the equation on page 106. Dirac is a poet when he writes physics: he simplifies everything in the extreme. "Noncommutative" means: such that their order cannot be changed freely. Dirac calls the noncommuting variables "q-numbers": quantities *defined* by this equation. Their pretentious mathematical name is "noncommutative algebra."

Remember Zeilinger's photons, with which I began to describe quantum phenomena? They could pass on the right or the left and end up either up or down. Their behavior may be described by two variables: a variable *X* that can have the value "right" or "left," and a variable *P* that can have the value "up" or "down." These two variables are like the position and speed of a particle: they do not commute. Hence they cannot be determined together. This is the reason why, if we close one of the paths determining the first variable ("right" or "left"), the second is undetermined: the photons go randomly "up" or "down." Vice versa, in order for the second variable to be determined, for the photons to all go "down," it is necessary that the first variable should not be determined; that is, that the photons must pass via both paths "right" and "left." The entire phenomenon follows from the equation which says that these two variables "do not commute" (are noncommutative), and hence cannot be determined together.

A single equation codes quantum theory. It implies that the world is not continuous but granular. There is no infinite in going toward the small: things cannot get infinitely smaller. It tells us that the future is not determined by the present. It tells us that physical things have properties only in relation to other physical things, and that these properties make sense only when things interact. It tells us that sometimes perspectives cannot be juxtaposed.

In our everyday life we are not aware of any of this. Quantum interference gets lost in the buzz of the macroscopic world. We can reveal it only through delicate observations, isolating objects as much as possible.[77]

If we do not observe interference, we can ignore superposition and reinterpret it as ignorance: we just don't know if the cat is asleep or awake. We have no need to think that there is a quantum superposition because *quantum superposition*—I emphasize it as there is often confusion on the issue—means *only* that we see interference. The delicate phenomena of interference between the cat awake and the cat sleeping are lost in the noise of the world that surrounds us. When interference is lost, we can take facts as *stable*, that is, we can forget that they are only true relative to something else.[78]

Furthermore, when we observe the world at our scale, we do not see its *granularity*. We cannot see single molecules: we see the whole cat. With many variables, fluctuations become irrelevant, and probability nears certainty.[79] Billions of discontinuous events of the agitated and fluctuating quantum world are reduced by us to the few continuous and well-defined variables of our everyday experience. At our scale, the world is like the wave-agitated surface of the ocean seen from the moon: the smooth surface of a blue marble.

Our everyday experience is thus compatible with the quantum world: quantum theory *incorporates* classical mechanics and our usual vision of the world—as approximations. We understand it as a man with good sight can understand the experience of a myopic person. But at the molecular scale, the cutting edge of a sharp knife is as fluctuating and imprecise as the edge of an ocean in a storm, fraying upon the white sand of its shore.

The solidity of the classical vision of the world is nothing other than our own myopia. The certainties of classical physics are just probabilities. The well-defined and solid picture of the world given by the old physics is an illusion.

⁂

On April 18, 1947, on the sacred island of Helgoland, the Royal Navy blows up three thousand nine hundred ninety-seven tons of dynamite—what was left of munitions abandoned there by the German army. It is probably the biggest explosion ever made using conventional explosives. Helgoland is devastated. It is almost as if humanity was seeking to cancel out the rip in reality opened on the island by a young physicist.

But the rip remains. The conceptual explosion unleashed by him is more devastating than anything thousands of tons of TNT could produce—it is the very framework of reality as we knew it that has shattered.

There is something disorienting in all this. The solid basis of reality seems to melt between our fingers, in an infinite regression of references.

I stop writing these lines and look out of the window. There is still snow. Here in Canada, spring comes late. In my room there is a fire in the fireplace. I get up and add another log to it. I am writing about the nature of reality. I look into the fire and wonder which reality I am speaking about. This snow? This flickering fire? Or the reality which I've read about in books? Perhaps only the warmth from the fire reaching my skin, the nameless reddish-orange flickers, the tenuous whitish-blue of approaching twilight?

For a moment even these sensations melt. I close my eyes and see bright lakes of vivid color parting like curtains, through which I feel I am falling. Is this also reality? Violet and orange shapes are dancing, and I am no longer there.

I take a sip of tea. Stoke the fire. Smile. We navigate in an uncertain sea of colors and have at our disposal good maps with which to orient ourselves. But between our mental maps and reality there is the same distance as between the charts of sailors and the fury of the waves crashing against the cliffs, where the gulls hover and cry.

That fragile web, our mental organization, is little more than a clumsy tool for navigating through the infinite mysteries of this magical light-flooded kaleidoscope in which we are amazed to exist and that we call our world.

We can traverse it unquestioningly, with faith in the maps that we have; after all, they allow us to live pretty well. We can remain quiet, overwhelmed by the light and by the infinite beauty. We can sit patiently at a desk, light a candle or turn on a MacBook Air, go to the laboratories, discuss with friends and enemies, retire to a Sacred Island to calculate and to clamber across rocks at dawn; or drink a little tea, revive the flames in the fireplace and start to write again, trying to understand together a few more grains of truth, to pick up that mariners' chart again and contribute to improving a bit of it. Once again, to rethink Nature.

PART THREE

V

THE UNAMBIGUOUS DESCRIPTION OF AN OBJECT INCLUDES THE OBJECTS TO WHICH IT MANIFESTS ITSELF

In which I ask what all of this means for our ideas about reality and realize that the conceptual novelty of the theory is not so new, after all.

ALEKSANDR BOGDANOV AND VLADIMIR LENIN

In 1909, four years after the failed Revolution of 1905 and eight years before the victorious October Revolution, Lenin, under the pseudonym "V. Il'in," published *Materialism and Empirio-Criticism: Critical Comments on a Reactionary Philosophy*, his most philosophical text.[80]

The implicit political target was Aleksandr Bogdanov, the cofounder and principal thinker of the Bolsheviks, until then Lenin's friend and ally.[81]

In the years preceding the revolution, Bogdanov had published a work in three volumes offering a general theoretical basis for the revolutionary movement.[82] It made reference to a philosophical perspective called "empiriocriticism." Lenin had begun to see in Bogdanov a serious rival and had come to fear his ideological influence. In his own book, Lenin ferociously criticizes empiriocriticism as "reactionary philosophy," and staunchly defends what he calls "materialism."

"Empiriocriticism" was the name Ernst Mach had associated to his own ideas. Ernst Mach—remember him?—was the source of philosophical inspiration for both Einstein and Heisenberg.

Mach is not a systematic philosopher; his work at times lacks clarity. And yet I believe that the extent and depth of his influence on contemporary culture has been undervalued.[83] Mach inspired the beginning of both of the great twentieth-century revolutions in physics: relativity and quantum theory. He played a direct role in the birth of the scientific study of perception. He was at the center of the politico-philosophical debate that led to the Russian Revolution. He had a determining influence on the founders of the Vienna Circle (the official name of which was Verein Ernst Mach, or the Ernst Mach Society), the philosophical environment that proved to be fertile ground for logical positivism, which directly inherits from Mach its "anti-metaphysical" rhetoric and is the root of so much contemporary philosophy of science. His direct influence reaches to American pragmatism, another root of today's analytic philosophy.

Mach even made a mark on literature. One of the outstanding novelists of the twentieth century, Robert Musil, wrote his doctoral thesis on Mach's work. The turbulent discussions engaged in by the protagonist of his first novel, *The Confusions of Young Törless* (1906), revisit the themes of the thesis on the meaning of the scientific reading of the world. Musil's major work, *The Man without Qualities* (1930–43), is also filigreed with such questions—from its very first page, which opens with a crafty double description, scientific and quotidian, of a sunny day.[84]

Mach's influence on the revolution in physics was almost personal. An old friend of Wolfgang Pauli's father, he was the godfather of the Pauli with whom Heisenberg would discuss philosophy. Einstein had, as a friend and fellow student in Zurich, Friedrich Adler, the son of a cofounder of the Austrian Social Democratic Workers' Party and the promoter of a convergence of ideas between Mach and Marx. Adler would become leader of the Social Democratic Workers' Party and, in protest against Austrian participation in the First World War, would later assassinate the country's prime minister, Karl von Stürgkh. While in prison, Adler would write a book on the subject of . . . Ernst Mach.[85]

In short, Mach stands at a remarkable crossroads of science, politics, philosophy and literature. And some people still view the natural sciences, humanities and literature as unconnected!

Mach's chief polemical target was eighteenth-century mechanism: the idea that all phenomena are products of matter that move through space. Mach argues that the progress of science shows that this notion of "matter" is an unjustified "metaphysical" assumption: a model that was useful for a time, but which we needed to move past, so that it does not become a metaphysical prejudice. Science must be freed from *all* metaphysical assumptions: knowledge should be based, that is, only upon what is "observable."

Does this remind you of anything? This is exactly the premise of Heisenberg's magical work conceived on the island of Helgoland—the work that opened the way to quantum theory and the story told in this book. Heisenberg's article begins thus: "It is the aim of this work to lay the foundation for a theory of quantum mechanics based solely on relations between quantities that are in principle observable." Almost a quotation from Mach.

The idea that knowledge must be founded on experience and observations is certainly not original: it is the classical tradition of empiricism that goes back to Locke and to Hume, if not all the way back to Aristotle. But the focus upon the relation between the subject and the object of knowledge, and doubts about the possibility of knowing the world "as it really is," had led to the great German idealists and the philosophical centrality of the subject who holds the knowledge. Mach, a scientist, diverts this attention from the subject to experience itself—to what he calls "sensations." He studies the concrete form in which scientific knowledge grows on the basis of experience. His best-known work examines the historical evolution of mechanics.[86] It interprets it as the effort to summarize in the most economical way the known facts on movement as revealed by sensations.

Mach does not see knowledge as a question of deducing or intuiting a hypothetical reality *beyond sensations*, but as the search for an efficient organization of our way of thinking about these sensations. The world that interests us, for Mach, *is constituted by* sensations. Any firm assumption about what lies "behind" those sensations is suspect as being a form of "metaphysics."

The notion of "sensation" is ambiguous in Mach. This is both his weakness and his strength: Mach takes the concept of sensation from physiology but makes it serve as a universal notion *independent from the psychological sphere*. He uses the term "elements" (in a sense similar to *dhamma* in Buddhist philosophy). "Elements" are not just the sensations that a human being or an animal experiences. They are any phenomena that manifest themselves in the universe. The "elements" are not independent: they are tied by relations, what Mach calls "functions," and these are what science studies. Though imprecise, Mach's philosophy is a real natural philosophy that replaces the mechanism of matter that moves in space with a general set of elements and functions.[87]

The appeal of this philosophical position is that it eliminates every firm hypothesis concerning a reality that exists behind appearances, but also every hypothesis on the reality of the subject who experiences. For Mach, there is no distinction between the physical and the mental world: "sensation" is equally physical and mental. It is real. Bertrand Russell describes the same idea thus: "the raw material out of which the world is built up is not of two sorts, one matter and the other mind; it is simply arranged in different patterns by its inter-relations: some arrangements may be called mental, while others may be called physical."[88] The idea of a material reality behind phenomena disappears; the idea of a spirit that "knows" disappears. Knowledge is possessed, for Mach, not by the abstract "subject" of idealism: it is instead the concrete human activity, in the concrete course of history, that learns to better and better organize the facts of the world with which it interacts.

This perspective, historical and concrete, resonates with the ideas of Marx and Engels, for whom knowledge is part of a concrete human history. Knowledge is divested of any ahistorical element, of every aspiration toward the absolute or pretense of certainty; it is located instead in the actual biological, historical and cultural evolution of mankind on our planet. It comes to be interpreted in terms of biology and economics, as a tool for simplifying our interaction with the world. It is not a definitive acquisition but an ongoing process. For Mach, knowledge is the science of nature, but its perspective is not far from the historicism of dialectical materialism. The consonance between Mach's ideas and those of Engels and Marx is developed by Bogdanov and gains currency in prerevolutionary Russia.

Lenin's response is scathing. In *Materialism and Empirio-Criticism* he launches an all-out attack upon Mach, his Russian disciples and, by implication, Bogdanov. He accuses them of that gravest of sins: the practice of "reactionary" philosophy. In 1909, Bogdanov is expelled from the editorial committee of *The Proletarian*, the underground newspaper of the Bolsheviks, and shortly after from the Central Committee of the party.

Lenin's critique of Mach and Bogdanov's reply interest us here.[89] Not because Lenin is Lenin, but because his criticism is the natural reaction to the ideas that led to quantum theory. The same criticism occurs naturally to us. Precisely the issues debated by Lenin and Bogdanov have returned in contemporary philosophy. Their discussion provides a key for understanding the revolutionary significance of quanta.

⁂

Lenin accuses Bogdanov and Mach of being "idealists." An idealist, for Lenin, negates the existence of a real world beyond the spirit and reduces reality to the content of the mind.

If only "sensations" are real, argues Lenin, then external reality is assumed not to exist: we live in a solipsistic world where there is only myself and my sensations. I take myself, the subject, as the only reality. This idealism, for Lenin, is the ideological manifestation of the enemy: it is pure bourgeois-ism. Against idealism, Lenin poses a materialism that sees the human being—human consciousness, human spirit—as an aspect of a concrete world that is objective, knowable, and comprising solely matter in motion in space.

Whatever we think about communism, there is no denying that Lenin was an extraordinary politician. His knowledge of philosophy is also impressive; if today we elected politicians as cultivated as Lenin, perhaps they would be more effective. But Lenin was no great philosopher. The influence of his philosophical writings is due more to his long dominance of the political scene, and his elevation to heroic status under Stalin, than to the profundity of his arguments. Mach deserves better.[90]

Bogdanov, indeed, replies to Lenin that his criticism misses the point. Mach's thought is not idealism, much less solipsism. The humanity that knows is not an isolated, transcendent subject; it is not the philosophical "I" of idealism: it is real humanity, immersed in concrete history, part of the natural world. The "sensations" are not "within our mind." They are natural phenomena in the world: the form in which the world presents itself to the world. They do not come to a self that is separate from the world: they come to the skin, to the brain, to the neurons of the retina, to the receptors in our ears. These are elements of nature.

Lenin defines "materialism" in his book as the belief that a world exists beyond our minds.[91] If this is the definition of materialism, then Mach is definitely a materialist; we are all materialists. Even the pope is a materialist. But then, for Lenin, the only acceptable version of materialism is the idea that "there is nothing in the world except matter in motion in space and in time," and that we can arrive at "absolute truths" through knowledge of matter. Bogdanov highlights the *scientific* as much as the *historical* weakness of these peremptory assertions. Of course the world is outside our mind, but things are much more subtle than naive materialism would have it. The choice is not just between the idea that the world exists only in our minds and the idea that it consists only of particles of matter in motion.

Mach does not think, of course, that there is nothing outside our mind. On the contrary, he is interested precisely in what is outside our minds (whatever the "mind" is): nature, in all its complexity, of which we are a part. Nature presents itself as a set of phenomena, and Mach recommends the study of those phenomena in order to build syntheses and conceptual structures that make sense of them, rather than to postulate a priori underlying realities.

His most radical suggestion is to stop thinking of phenomena as manifestations of objects and to think, instead, of objects as nodes between phenomena. This is not a metaphysics of the contents of consciousness, as

Lenin sees it: it is a step back with regard to the metaphysics of the objects-in-themselves. Mach is witheringly dismissive: "The conception of the [mechanistic] world appears to us to be mechanical mythology, like the animistic mythology of ancient religions."[92]

Einstein recognized his debt to Mach on numerous occasions.[93] The critique of the ("metaphysical") assumption of the existence of a real fixed space "within which" things move opened the doors to his general relativity.

In the space opened up by Mach's reading of science—which does not take the reality of anything for granted, except to the extent that it allows phenomena to be organized—Heisenberg slips in, to remove from the electron its trajectory and to reinterpret it solely in terms of its manifestations.

In this same space, the possibility of a *relational* interpretation of quantum mechanics opens up: the elements useful for thinking the world are manifestations of physical systems to each other, not absolute properties belonging to each system.

Bogdanov criticizes Lenin for making "matter" an absolute and ahistorical category, a "metaphysical" one in the sense given to the word by Mach. He disapproves, above all, of Lenin's forgetting one of the essential lessons of Marx and Engels: History is process, knowledge is process. Scientific knowledge grows, writes Bogdanov, and the notion of matter proper to the science of our time may turn out to be only an intermediate stage on the path of our knowledge. Reality may be much more complex than the naive materialism of eighteenth-century physics. Prophetic words, for just a few years later Werner Heisenberg would open the door to the quantum level of reality.

Even more impressive is Bogdanov's *political* reply to Lenin. Lenin speaks of absolute certainties. He presents the historical materialism of Marx and Engels as if it were timelessly valid. Bogdanov points out that this ideological dogmatism not only fails to accord with the dynamic of scientific thought, it also leads to calcified political dogmatism. The Russian Revolution, Bogdanov argues in the turbulent years of its aftermath, had created a new economic structure. If, as Marx suggested, culture is influenced by economic structure, then post-revolutionary society would be able to produce a new culture that could no longer be the orthodox Marxism conceived *before* the revolution. Brilliant. Bogdanov's political program was to leave power and culture to the people, to nurture the new, collective, generous culture opened up by the revolutionary dream. Lenin's political program, instead, was to reinforce the revolutionary avant-garde, the repository of Truth that needed to *guide* the people. The rather repulsive style in which *Materialism and Empirio-Criticism* is written reflects its philosophical stance: it has been called "an angry moral tone, dim echo of anathema and excommunication"—"possibly the rudest work of philosophy ever published."[94]

Bogdanov predicts that Lenin's dogmatism would seal the Russian Revolution into a block of ice; prevent it from evolving further; suffocate the life out of all that had been gained through it; render it sclerotic. These, too, were prophetic words.

⁂

"Bogdanov" is a pseudonym; one of the many that he used to hide from the tsar's police. He was born Aleksandr Aleksandrovič Malinovskij, the son of a village schoolteacher and the second of six brothers. Independent and rebellious from a very early age, legend has it that the first words he spoke, at eighteen months during a family squabble, were "Dad's an idiot!"[95]

Thanks to the promotion of his father (who was not an idiot) to a position teaching physics in a city with larger schools, young Aleksandr has access to a library and a rudimentary physics lab. He gets a scholarship to attend high school, about which he would later write that "the mental closure and the malice of the professors taught me to be wary of the powers that be, and to resist all authority."[96] The same visceral dislike of authority guided the development of his slightly younger contemporary Albert Einstein.

Having graduated brilliantly from school, Bogdanov enrolls at Moscow University to study natural sciences. He joins a student organization that helps comrades from distant provinces. He becomes involved in political activities. He is arrested multiple times. He contributes to the publication of Karl Marx's *Das Kapital* and Mach's *Analysis of Sensations* in Russia. He works in political propaganda, writes popularizing texts on economics for the workers. He studies medicine in Ukraine, gets arrested again and exiled. In Zurich he becomes acquainted with Lenin. He is a leader of the Bolshevik movement, something like Lenin's deputy leader, in fact. In the following years, after the arguments with Lenin, he becomes distanced from the leadership, and after the revolution is kept at a distance from the centers of power. He remains universally respected and continues to exert a strong cultural, moral and political influence. In the 1920s and 1930s, he is a reference point for the underground "left-wing" opposition that seeks to defend the successes of the revolution from Bolshevik autocracy, until this dissidence is crushed by Stalin.

The key concept of Bogdanov's theoretical work is the notion of *organization*. Social life is the organization of collective work. Knowledge is the organization of experience and of concepts. It is possible to understand the whole of reality as organization, structure. The picture of the world that Bogdanov proposes is based on a spectrum of kinds of organization that become gradually more complex: from minimal elements that interact directly, through the organization of matter in the living, the biological development of individual experience organized in individuals, up to scientific knowledge, which, for Bogdanov, is collectively organized experience. Through the cybernetics of Norbert Wiener and the system theory of Ludwig von Bertalanffy, these ideas will have a little-recognized but profound influence on modern thought, on the birth of cybernetics, on the science of complex systems, down to contemporary structural realism.

In Soviet Russia, Bogdanov becomes a professor of economics at Moscow University, directs the Communist Academy and republishes his early sci-fi novel *Red Star*, which becomes a huge publishing success. The novel describes a utopian, libertarian society on Mars that has overcome all distinctions between men and women and that uses an efficient statistical apparatus for processing economic data that can indicate to industries exactly what needs to be produced, and to the unemployed precisely in which factories to seek work, and so on, leaving everyone the freedom to choose how they should live.

Bogdanov focuses on organizing centers for proletarian culture, where a novel culture, based on collaboration, mutually supportive rather than competitive, can flower autonomously. Moved on from this activity by Lenin, he devotes himself to medicine. A doctor by training, he had served as such at the front during the First World War. He founds an institute for medical research in Moscow and becomes one of the pioneers of blood transfusion. In his revolutionary and collectivist ideology, the transfusion of blood is symbolic of the potential for men and women to collaborate and share.

A doctor, an economist, a philosopher, a natural scientist, a science-fiction novelist, a poet, teacher, politician, progenitor of cybernetics and of the science of organization, a pioneer of blood transfusion and a lifelong revolutionary, Aleksandr Bogdanov, prodigiously talented,[97] is one of the most complex and fascinating figures of the intellectual world at the beginning of the twentieth century. His ideas, too radical for both sides of the Iron Curtain, have spread slowly, in a subterranean way. Only very recently has the three-volume work that gave rise to Lenin's critique been published in English translation. Curiously, we find more traces of Bogdanov in literary works: as the inspiration for the novel *Proletkult* by Wu Ming, and for the great character Arkady Bogdanov in Kim Stanley Robinson's delightful trilogy *Red Mars*, *Green Mars*, *Blue Mars*.[98]

Faithful to the end to his ideal of sharing, Aleksandr Bogdanov dies an incredible death, in a scientific experiment in which he exchanges his blood with a young man suffering from tuberculosis and malaria in an attempt to cure him.

Right up until the end, the courage to experiment, the courage to exchange and share, the dream—and the practice—of brotherhood.

NATURALISM WITHOUT SUBSTANCE: CONTEXTUALITY

I have digressed. But it is the perspective provided by Mach that allowed Heisenberg to take his crucial step, and the polemic between Lenin and Bogdanov highlights the issue that generates misunderstandings around quantum theory.

The "anti-metaphysical" spirit that Mach promoted is an attitude of openness: We should not seek to teach the world how it should be. Let's listen to the world instead, in order to learn from it how to think about it.

When Einstein objected to quantum mechanics by remarking that "God does not play dice," Bohr responded by admonishing him, "Stop telling God what to do." Which means: Nature is richer than our metaphysical prejudices. It has more imagination than we do.

One of the most acute philosophers to have examined quantum theory, David Albert, once asked me:

"Carlo, how can you think that experiments in a laboratory made with little bits of metal and glass can have such significance as to put into question our most rooted metaphysical convictions about how the world works?" The question has haunted me ever since. But in the end the answer seems simple to me: "What are these 'most rooted metaphysical convictions' of ours, if not what we have become accustomed to believe precisely by handling stones and pieces of wood?"

Our prejudices concerning how reality is made are just the result of our experience. Our experience is limited. We cannot take as gospel truth the generalizations that we have made in the past. No one puts this better than Douglas Adams, author of *The Hitchhiker's Guide to the Galaxy*, with his characteristic mordancy: "The fact that we live at the bottom of a deep gravity well, on the surface of a gas covered planet going around a nuclear fireball 90 million miles away and think this to be normal is obviously some indication of how skewed our perspective tends to be."[99]

We must expect to have to modify our provincial metaphysical perspectives just as soon as we learn something new. We must take seriously the new things we learn about the world, even if they clash with our preconceptions about how reality is constituted.

This seems to me an attitude that renounces the arrogance of possessing knowledge, while keeping faith with reason and our capacity to learn. Science is not a Depository of Truth, it is based on the awareness that *there are no* Depositories of Truth. The best way to learn is to interact with the world while seeking to understand it, readjusting our mental schemes to what we encounter and find. This respect for science as the source of our knowledge nurtures the naturalism of philosophers such as Willard Quine, for whom our knowledge itself is one of many natural processes and can be studied as such.

Many "interpretations" of quantum mechanics, like those referred to in Chapter II, seem to me to be efforts to squeeze the discoveries of quantum physics into the canons of metaphysical prejudice. Are we convinced that the world is deterministic, that the future and the past are univocally determined by the world's present? Then let's add quantities that determine the past and the future, even if they are unobservable. Does it disturb us to see a component of a quantum superposition disappear? Then let's introduce a parallel universe where this component can go and hide. And so on.

I believe that we need to adapt our philosophy to our science, and not our science to our philosophy.

⁂

Niels Bohr was the spiritual father of the young Turks who built quantum theory. He pushed Heisenberg to preoccupy himself with the problem and accompanied him on his delvings into the mysteries of atoms. He mediated the argument between Heisenberg and Schrödinger, his two overly brilliant bickering children. It was he who formulated the way of thinking about the theory that has ended up in physics books all over the world. He was the scientist who perhaps strained more than any other to understand what it all meant. His discussion with Einstein on the plausibility of the theory lasted for years, forcing both these giants to clarify their positions, and to compromise.

Einstein had always recognized that quantum mechanics was a significant step forward in our comprehension of the world: it was he who proposed Heisenberg, Born and Jordan for the Nobel Prize. But he was never convinced by the form that the theory had taken. He accused it, at different times, of being inconsistent, implausible and incomplete.

Bohr defended the theory from Einstein's criticisms, sometimes soundly, sometimes even winning discussions with arguments that were wrong.[100] Bohr's thought is not limpid; in fact, it is often somewhat obscure. But his intuitions are extremely acute and helped to build a good part of our current understanding of the theory. His key intuition is summarized in this observation:

> *Whereas in classical physics the interactions between an object and the measuring apparatus can be overlooked—or if necessary can be taken into account and compensated for—in quantum physics this interaction is an inseparable part of the phenomenon. For this reason, the unambiguous description of a quantum phenomenon is required in principle to include a description of all the relevant aspects of the experimental arrangement.*[101]

These words capture the relational aspect of quantum mechanics, but in the circumscribed context of a phenomenon observed in a laboratory with instruments of measurement. For this reason, they lend themselves to a misunderstanding: the idea that we are speaking only of a situation in which there is a particular observer, who uses instruments of measurement. To think that a human being, their mind, their tools or the numbers they use plays any special role in the grammar of nature is nonsense.

What we need to add to Bohr's paragraph is the awareness, which has grown in the course of a century of successes for the theory, of the fact that *all* nature is quantum, and that there is nothing special about a physics laboratory containing measuring apparatus. There are not quantum phenomena only in laboratories and non-quantum phenomena elsewhere: all phenomena are quantum phenomena. Extended to any and every natural phenomenon, Bohr's intuition becomes:

> *Whereas previously we thought that the properties of every object could be determined even if we overlooked the interactions occurring between this object and others, quantum physics demonstrates that the interaction is an inseparable part of phenomena. The unambiguous description of any phenomenon requires the inclusion of all the objects involved in the interaction in which the phenomenon manifests itself.*

Now, *this* is radical, but it is clear. Phenomena are actions by one part of the natural world upon another part of the natural world. Confusing this discovery with something that has to do with our minds is the mistake Lenin made: in the argument with Mach, Lenin is the dualist—the one, in other words, who is unable to conceive of phenomena if not relative to a transcendent subject.

The mind does not enter into the equation. Special "observers" have no real role to play in the theory. The central point is simpler: the properties of an object become manifest when this object interacts with others. We cannot separate the properties from these other objects. We cannot attribute them just to a single object. All of the (variable) properties of an object, in the final analysis, are such and exist only with respect to other objects.

"Contextuality" is the technical name that denotes this central aspect of quantum physics: things exist in a context.

An isolated object, taken in itself, independent of every interaction, has no particular state. At most we can attribute to it a kind of probabilistic disposition to manifest itself in one way or another.[102] But even this is only an anticipation of future phenomena, a reflection of phenomena past, and only and always relative to another object.

The conclusion is revolutionary. It leaps beyond the idea that the world is made up of a substance that has attributes, and forces us to think about everything in terms of relations.[103]

This, I believe, is what we have discovered about the world with quanta.

WITHOUT FOUNDATION? NĀGĀRJUNA

This way of understanding the central discovery of quantum mechanics is rooted in the original intuitions of Heisenberg and Bohr, but was formalized in the mid-1990s with the birth of the "relational interpretation of quantum mechanics."[104] The world of philosophy has reacted to this interpretation in various ways: different schools of thought have framed it in different philosophical terms.

Bas van Fraassen, one of the most brilliant contemporary philosophers, gave it an acute analysis within the framework of his "constructive empiricism."[105] Michel

Bitbol gave it a neo-Kantian reading.[106] François-Igor Pris read it in the perspective of contextual realism; Pierre Livet in terms of the ontology of processes.[107] Mauro Dorato has inserted it into structural realism, according to which reality is made up of structures.[108] Laura Candiotto has defended the same thesis.[109] I do not intend to enter into the debate among different currents of contemporary philosophy. I only add here a few pointers, and tell a personal story.

The discovery that quantities we had thought of as absolute are, in fact, relative instead is a theme that runs throughout the history of physics. Beyond physics, relational thinking can be found in all the sciences. In biology, the characteristics of living systems are comprehensible in relation to their environment formed by other living beings. In chemistry, the properties of elements consist of the way in which they interact with other elements. In economics, we speak of economic relations. In psychology, the individual personality exists within a relational context. In these and many other cases, we understand things (organisms, chemicals, psychological life) through their being *in relation to* other things.

In the history of Western philosophy, there is a recurrent critique of the notion that "entities" are the foundation of reality. It can be found in widely different philosophical traditions, from the "Everything flows" of Heraclitus to the contemporary metaphysics of relations.[110] Only in the past few years, books of philosophy have come out with titles such as *Formal Approach to the Metaphysics of Perspectives* and *Viewpoint Relativism: A New Approach to Epistemological Relativism Based on the Concept of Points of View*, to name just some most recent examples.[111] In analytic philosophy, structural realism is based on the idea that relations come before objects.[112] Michel Bitbol has written *From inside the World: For a Philosophy and a Science of Relations*.[113] Laura Candiotto and Giacomo Pezzano have published a book with the title *The Philosophy of Relations*.[114]

But the idea itself is ancient. In the Western tradition, we can already find it in the later work of Plato. In the *Sophist*, Plato considers the fact that his atemporal Forms must be able to enter into relation with phenomenal reality to make sense, and ends up putting into the mouth of the central figure in his dialogue, the Stranger from Elea, a famous, completely relational (and not very Eleatic) definition of reality: "I say therefore that what by nature can act on another or suffer even the slightest action from another, however insignificant it is, and even if it happens only once, this alone can be truly real. I therefore propose this definition of being: that it is nothing if not action (δύναμις)."[115] As is not uncommon, someone might be tempted to think that Plato has summed up in a phrase everything that there is to be said on the subject . . .

Even this very incomplete overview is sufficient to show how recurrent is the idea that the world is woven by relations and interactions more than by objects.

Take an object: this chair that I see in front of me. It is real and stands before me, objectively, no doubt about it. But what does it mean, exactly, that this whole is an object, an entity, a chair, real?

The notion of a chair is defined by its function: a piece of furniture designed for us to sit on. It presupposes human beings who sit down. It's about the way we conceive of it.

This does not affect the fact that the chair exists right here, objectively. The object is still here, with its obvious physical characteristics of color, hardness and so on. But even these characteristics exist only in relation to us. Color comes from the encounter between the frequencies of light reflected from the surfaces of the chair and the particular receptors in human retinas. It is not about the chair: it is a story between light, retina and reflection. Most other animal species do not see colors as we do. The frequencies themselves emitted by the chair emerge only from the interaction between the dynamics of its atoms and the light that illuminates them.

The chair, still, is an object independent of its color. If I move it, it moves as a whole. Strictly speaking, not even this is completely true: this chair is made of a seat that rests on a frame, which rises when I pick it up. It is a set, an assemblage of pieces.

What is it that makes this assemblage of pieces a single object, a unit? Effectively, it is little more than the role that this combination of elements plays for us.

If we look for the chair in itself, independently of external relations, and especially of its relations to us, we struggle to find it.

There is nothing mysterious about this: the world is not divided into stand-alone entities. It is we who divide it into objects for our convenience. A mountain chain is not divided into individual mountains: it is we who divide it up into parts that strike us as in some way separate. A countless number of our definitions, perhaps all of them, are relational: a mother is a mother because she has a child; a planet is a planet because it orbits a star; a predator is such because it hunts prey; a position in space is there only in relation to something else. Even time exists only as a set of relations.[116]

None of this is new. But physics has long been asked to provide a firm basis on which to place relations: a basic reality underlying and supporting this relational world. Classical physics, with its idea of matter that moves in space, characterized by primary qualities (shape) that come before secondary ones (color), seemed to be able to play this role. It appeared to furnish the primary ingredients of the world that it was possible to think of as existing in their own right as the basis of the interplay of combinations and relations.

The discovery of the quantum properties of the world is the discovery that physical matter is not capable of fulfilling this role. Fundamental physics does provide an elementary and universal grammar for understanding phenomena, but not a grammar consisting of simple matter in motion, with its own primary properties. The contextuality that permeates the world reaches this elementary grammar. There are no elementary entities that we can describe except in the context of their interaction with something else.

This leaves us without a foothold, no place to stand. If matter with definite and univocal properties does not constitute the elementary substance of the world, and if the subject of our knowledge is a part of nature, what *is* the world's elementary substance?

To what can we anchor our conception of the world? From where can we begin? What is fundamental?

The history of Western philosophy is to a large extent an attempt to provide an answer to the question as to what is fundamental. It is a search for the point of departure from which everything else follows: matter, God, the spirit, the atoms and the void, Platonic Forms, a priori forms of intuition, the subject, Absolute Spirit, elementary moments of consciousness, phenomena, energy, experience, sensations, language, verifiable propositions, scientific data, falsifiable theories, the existence of the being for whom being matters, hermeneutic circles, structures . . . A long list of candidates, not one of which ever managed to achieve a universal acceptance as ultimate foundation.

The attempt by Mach to take "sensations" or "elements" as foundational has inspired scientists and philosophers, but in the end does not seem any more convincing than others. Mach rails against metaphysics, but he effectively produces his own metaphysics—lighter and more flexible, but a metaphysics nonetheless—of elements and functions. Mach's is a phenomenal realism, or a "realist empiricism."[117]

In my own attempts to make sense of quanta for myself, I have wandered among the texts of philosophers in search of a conceptual basis with which to understand the strange picture of the world provided by this incredible theory. In doing so, I have found many fine suggestions and acute criticisms, but nothing wholly convincing.

Until one day I came across a work that left me amazed. I will end this chapter, which does not have any conclusions, with a light account of this encounter.

I did not come across it by chance. When speaking about quanta and their relational nature, I had frequently met people who asked: Have you read Nāgārjuna?

When I'd heard my umpteenth "Have you read

Nāgārjuna?" I decided to go ahead and read it. Though not widely known in the West, the work in question is hardly an obscure or minor one: it is one of the most important texts of Buddhist philosophy, so it was only due to my personal ignorance of Asian thought (not so uncharacteristic in the West) that I was unaware of it. Its title is one of those never-ending Sanskrit words—*Mūlamadhyamakakārikā*—translated in numerous ways, including *The Fundamental Wisdom of the Middle Way*. I read it in a translation with commentary by an American analytic philosopher. It has made a profound impression upon me.[118]

Nāgārjuna lived in the second century CE. There have been countless commentaries on his text, which has been overlaid with interpretations and exegesis. The interest of such ancient texts lies partly in the stratification of readings that gives them to us enriched by levels of meaning. What really interests us about ancient texts is not what the author initially intended to say: it is how the work can speak to us now, and what it can suggest today.

The central thesis of Nāgārjuna's book is simply that there is nothing that exists in itself independently from something else. The resonance with quantum mechanics is immediate. Obviously, Nāgārjuna knew nothing, and could not have imagined anything, about quanta—that is not the point. The point is that philosophers offer original ways of rethinking the world, and we can employ them if they turn out to be useful. The perspective offered by Nāgārjuna may perhaps make it a little easier to think about the quantum world.

If nothing exists in itself, everything exists only through dependence on something else, in relation to something else. The technical term used by Nāgārjuna to describe the absence of independent existence is "emptiness" (*śūnyatā*): things are "empty" in the sense of having no autonomous existence. They exist thanks to, as a function of, with respect to, in the perspective of, something else.

If I look at a cloudy sky—to take a simplistic example—I can see a castle and a dragon. Do a castle and a dragon really exist up there in the sky? Obviously not: the dragon and the castle emerge from the encounter between the shape of the clouds and the sensations and thoughts in my head; in themselves, they are empty entities, they do not exist. So far, so easy. But Nāgārjuna also suggests that the clouds, the sky, sensations, thoughts and my own head are equally things that arise from the encounter with other things: they are empty entities.

And myself, looking at a star, do I exist? No, not even I. So who is observing the star? No one, says Nāgārjuna. To see a star is a component of that set of interactions that I conventionally call my "self." "What articulates language does not exist. The circle of thoughts does not exist."[119] There is no ultimate or mysterious essence to understand—that is the true essence of our being. "I" is nothing other than the vast and interconnected set of phenomena that constitute it, each one dependent on something else. Centuries of Western speculation on the subject, and on the nature of consciousness, vanish like morning mist.

Like much philosophy and much science, Nāgārjuna distinguishes between two levels: conventional, apparent reality with its illusory and perspectival aspects, and ultimate reality. But in this case the distinction takes us in an unexpected direction: the ultimate reality, the essence, is absence, is vacuity. It does not exist.

If every metaphysics seeks a primary substance, an essence on which everything may depend, the point of departure from which everything follows, Nāgārjuna suggests that the ultimate substance, the point of departure . . . does not exist.

There are timid intuitions in a similar direction in Western philosophy. But Nāgārjuna's perspective is radical. Conventional, everyday existence is not negated; on the contrary, it is taken into account in all of its complexity, with its levels and facets. It can be studied, explored, analyzed, reduced to more elementary terms. But there is no sense, Nāgārjuna argues, in looking for an ultimate substratum.

The difference from contemporary structural realism, for instance, seems clear: I can imagine Nāgārjuna adding a short chapter to a contemporary edition of his book entitled "All Structures are Empty." They exist only when you are thinking about organizing something else. In his terms: "They are neither precedent to objects; nor not precedent to objects; neither are they both things; nor, ultimately, neither one nor the other thing."*

The illusoriness of the world, its *samsāra*, is a general theme of Buddhism; to recognize this is to reach *nirvāna*,

*This is an example of tetralemma, the form of logic used by Nāgārjuna.

liberation and beatitude. For Nāgārjuna, *samsāra* and *nirvāna* are the same thing: both empty of their own existence. Nonexistent.

So is emptiness the only reality? Is this, after all, the *ultimate* reality? No, writes Nāgārjuna, in the most vertiginous chapter of his book: every perspective exists only in interdependence with something else, there is never an ultimate reality—and this is the case for his own perspective as well. Even emptiness is devoid of essence: it is conventional. No metaphysics survives. Emptiness is empty.

Nāgārjuna has given us a formidable conceptual tool for thinking about the relationality of quanta: we can think of interdependence without autonomous essence entering the equation. In fact, interdependence—and this is the key argument made by Nāgārjuna—*requires* us to forget all about autonomous essences.

The long search for the "ultimate substance" in physics has passed through matter, molecules, atoms, fields, elementary particles . . . and has been shipwrecked in the relational complexity of quantum field theory and general relativity. Is it possible that a philosopher from ancient India can provide us with a conceptual tool with which to extricate ourselves?

ꕥ

It is always from others that we learn, from those different from ourselves. Despite millennia of uninterrupted dialogue, the East and the West still have something to say to each other. As in the best marriages.

The fascination of Nāgārjuna's thought goes beyond questions raised by contemporary physics. His perspective has something dizzying about it. It resonates with the best of much Western philosophy, both classical and recent. With the radical skepticism of Hume, with the unmasking of badly posed questions in Wittgenstein. But it seems to me that Nāgārjuna does not fall into the trap in which so much philosophy is caught, by postulating starting points that invariably turn out to be unconvincing in the long run. He speaks about reality, about its complexity and comprehensibility, but he defends us from the conceptual trap of wanting to find it an ultimate foundation.

His is not metaphysical extravagance: it is sobriety. It recognizes the fact that to inquire about the ultimate foundation of everything is to ask a question that perhaps simply does not make sense.

This does not shut down investigation. On the

contrary, it liberates it. Nāgārjuna is not a nihilist negating the reality of the world, and neither is he a skeptic denying that we can know anything about that reality. The world of phenomena is one that we can investigate, gradually improving our understanding of it. We may find general characteristics. But it is a world of interdependence and contingencies, not a world we should trouble ourselves attempting to derive from an Absolute.

I believe that one of the greatest mistakes made by human beings is to want certainties when trying to understand something. The search for knowledge is not nourished by certainty: it is nourished by a radical absence of certainty. Thanks to the acute awareness of our ignorance, we are open to doubt and can continue to learn and to learn better. This has always been the strength of scientific thinking—thinking born of curiosity, revolt, change. There is no cardinal or final fixed point, philosophical or methodological, with which to anchor the adventure of knowledge.

There are many different interpretations of Nāgārjuna's text. The multiplicity of potential readings is testimony to its vitality and to the capacity of ancient texts to continue to speak to us. What interests us, anew, is not what the prior of a monastery in India was actually thinking nearly two thousand years ago—that is his business (or the business of historians). What interests us is the power of the ideas that emanate today from the lines he left; how these, enriched by generations of commentary, may open up new spaces for thought, intersecting with *our* culture and *our* knowledge. This is the meaning of culture: an endless dialogue that enriches us by feeding on experiences, knowledge and, above all, exchanges.

I am not a philosopher, I am a physicist: a simple mechanic. And this simple mechanic, who deals with quanta, is taught by Nāgārjuna that it is possible to think of the manifestations of objects without having to ask what the object is *in itself,* independent from its manifestations.

But Nāgārjuna's emptiness also nourishes an ethical stance that clears the sky from the endless disquietude: to understand that we do not exist as autonomous entities helps us free ourselves from attachments and suffering. Precisely because of its impermanence, because of the absence of any absolute, the now has meaning and is precious.

For me as a human being, Nāgārjuna teaches the serenity, the lightness and the shining beauty of the world: we are nothing but images of images. Reality, including our selves, is nothing but a thin and fragile veil, beyond which . . . there is nothing.

VI

"FOR NATURE IT IS A PROBLEM ALREADY SOLVED"

In which I dare to ask myself where thoughts dwell. And whether the new physics could change a little the terms of this vexata quaestio.

SIMPLE MATTER?

> However mysterious the mind–body problem may be for us, we should always remember that it is a solved problem for nature.[120]

It is with sadness that every so often I spend a few hours on the internet, reading or listening to the mountain of stupidity dressed up with the word "quantum." Quantum medicine; holistic quantum theories of every kind;

mystical quantum spiritualism—and so on and on, in an almost unbelievable parade of quantum nonsense.

Worst of all is the pseudo-medicine. Every so often I receive an alarmed email from a relative of one of its victims: "My sister is being treated by a quantum medic. What do you think of it, Professor?" I think the worst it is possible to think: try to rescue your sister immediately. When it comes to medicine, this is the kind of situation in which the law, I believe, should be involved. Everyone has the right to seek to cure themselves as they see fit. But none has the right to cheat their fellow citizens with the kind of quackery that can cost lives.

Someone else writes me: "I have the sensation of having already lived this moment before: is this a quantum effect?" For pity's sake, no! What does the complexity of our memory and of our thoughts have to do with quanta? Absolutely nothing. Quantum mechanics has nothing to say about paranormal phenomena, alternative medicine or the influence of mysterious waves or vibrations.

For heaven's sake, I am all in favor of good vibrations. I, too, once had long hair tied with a red ban-

danna, and sat cross-legged next to Allen Ginsberg chanting "Om." But the delicate complexity of the emotional connection between ourselves and the universe has as much to do with ψ waves as a Bach cantata has to do with the carburetor in my old car.

The world is sufficiently complex to account for the beauty of Bach's music and the good vibrations of our deepest spiritual life, without the need to resort to the strangeness of quanta.

Or vice versa, if you will: the reality of quanta is *much stranger* than all the delicate, mysterious, enchanting, intricate aspects of our psychological reality and spiritual life. I find attempts to use quantum mechanics to explain complex phenomena that we know relatively little about, such as how the mind works, utterly unconvincing.

ꟗ

And yet, even if remote from our direct everyday experience, the discovery of the nature of the quantum world is too radical to have no relevance to precisely such big, open questions as the nature of mind. Not because the mind or other phenomena that we still know little

about are quantum phenomena, but because by modifying our conception of the physical world and of matter, the discovery of quanta changes the terms of our questions.

The conviction on which this book is based is that we human beings are a part of nature. We are a particular case among innumerable natural phenomena, none of which can escape the great laws of nature that are known to us. Yet who has never wondered, in some form or other: "If the world is made of simple matter, particles in motion in space, how is it that thoughts, subjectivity, values, beauty, meaning . . . come about?" How can "simple matter" produce colors, emotions and the burning lively sensation of existing? How is it that we can know and learn, be moved and amazed, read a book, understand, and even come to question how matter itself works?

Quantum mechanics doesn't have any direct answers to these questions. I fail to see any quantum explanation for subjectivity, perceptions, intelligence, consciousness or any other aspects of our mental life. Quantum phenomena intervene in the dynamics of atoms, photons, electromagnetic impulses and all other microscopic

structures which give rise to our body—but there is nothing specifically quantum that could help us understand what thoughts, perception and subjectivity are. These are aspects that involve the functioning of the brain at a large scale: that is precisely where quantum interference is lost in the noise of complexity. Quantum theory is of no direct help in understanding the mind.

But *indirectly* it may teach us something relevant, because it alters the terms of the problem.

It teaches us that the source of the confusion might lie in our erroneous intuitions, not just about the nature of consciousness (where our intuitions are certainly misleading), but also, crucially, about what "simple matter" is and how it functions.

Perhaps it is difficult to imagine how we as human beings may be made *only* of tiny stones bouncing against each other. But looked at closely, a stone is a vast world: a galaxy of swarming quantum entities where probabilities and interactions fluctuate. What we call "stone," furthermore, is a stratification of meanings in our thoughts, evoked by the interaction between ourselves and that galaxy of punctiform relative physical events. "Simple matter" splits apart into complex strata and suddenly seems much less simple. And perhaps the gulf between simple matter and the evanescent unraveling of our spirit appears to be a little less impassable.

If the fine grain of the world is made of material particles that have just mass and motion, it seems difficult to reconstruct our perceiving and thinking complexity from this amorphous grain. But if the fine grain of the world is better described in terms of relations, if nothing has intrinsic properties except in relation to other things, perhaps in *this* physics we can better find elements able to combine in a comprehensible way, to be the basis of those complex phenomena that we call our perceptions and our consciousness. If the physical world is woven from the subtle interplay of images in mirrors reflected in other mirrors, without the metaphysical foundation of a material substance, perhaps it becomes easier to recognize ourselves as part of that whole.

ꕥ

Someone has suggested that there is something psychic in all things. The argument is that since we have consciousness and are made up of protons and electrons, then the electrons and protons should already have a kind of proto-consciousness.

I don't find such arguments and such "pan-psychism" persuasive in the slightest. It is like saying that since a bicycle is made of atoms, then each atom must be proto-cyclist. Our mental life needs the existence of neurons, sensory organs, a body, the complex elaboration of information that occurs in our brains; all the evidence suggests that without this we have no mental life.

But there is no need to attribute proto-consciousness to elementary systems in order to get around a frozen "simple matter." It is enough to have observed how the world is better described by relative variables and their correlations. This allows us to be released from the prison of a blunt opposition between the objectivity of matter and mental life. The rigid distinction between a mental world and a physical one fades. It is possible to think of both mental and physical phenomena as natural phenomena: both products of interactions between parts of the physical world.

In this, the last chapter of the book before its conclusion, I offer a few humble suggestions in this difficult direction.

WHAT DOES "MEANING" MEAN?

We human critters live in a world of meanings. The words of our language "mean" something. The word "cat" means a cat. Our thoughts "signify"; they occur in our brain, but if we think of a tiger, we are referring to something that is not in our brain: the tiger may be anywhere out there in the world. If you are reading this book, you see the image of the lines on the page or screen. "To see" is something that happens in your brain, and yet the lines seen are "out there." A process takes place in your brain that refers to the lines on the paper. These, in turn, have a meaning: they refer to my thoughts while writing, which, in turn, refer to the you who is reading whom I am now imagining . . .

A technical term for "referring to something" in our mental processes (promoted by the German philosopher and psychologist Franz Brentano) is "intentionality." Intentionality is an important aspect of the notion of meaning and our whole mental life. There is a close relationship between what happens in thoughts and what happens "outside" of thoughts: what thoughts mean. There is a close relation between the word "cat" and a cat; between a road sign and what the road sign *signifies*.

There seems to be nothing of any of this in the natural world. A physical event in itself "means" nothing. A comet travels respecting the laws of Newton, but it does not do so by reading road signs . . .

If we are part of the physical world, this world of meanings must emerge from the physical world. How? What is the world of meanings, in purely physical terms?

Two concepts bring us close to an answer: *information* and *evolution*, even if neither is enough to comprehend what "meaning" is in physical terms. Let's consider them both.

ꝥ

In the information theory of Claude Shannon, *information* is only counting the number of possible states of something. A USB memory stick has a quantity of information, expressed in bits or gigabytes, which indicates how many different ways its memory can be arranged. The number of bits does not know the *meaning* of what is in the memory; it does not even know whether the content of the memory *means* something or is just noise.

Shannon also defines the notion of *relative information*, which is the one I used in the previous chapters: a measure of the physical correlation between two variables. Two variables have "relative information" if they can be in fewer states than the product of the number of states that each can be in.

This notion of "relative information" is purely physical. It is central in quantum physics: relative information is generated by the interactions that weave the world. Notice that it connects two different things, just as meaning does. But it isn't enough to understand meaning; the world swarms with correlations, but the vast majority of these do not signify or *mean* anything. To understand meaning we need something else.

The discovery of biological *evolution*, on the other hand, has allowed us to build some bridges between concepts that we use when speaking about animate things and concepts that we use for the rest of nature. In particular, it has clarified the biological—and in the final analysis, physical—origin of such notions as *utility* and *relevance*.

The biosphere is formed by structures and processes that are *useful* for the continuation of life: we have lungs *in order to* breathe; eyes *with which* to see. Darwin's discovery is that we understand why there are these structures by reversing the order of cause and effect between their utility and their existence: their function (to see, to eat, to breathe, to digest . . . to contribute to life) is not the *purpose* of these structures. It is the other way around: living beings survive *because* these structures are there. We do not love in order to live: we live because we love.

Life is a biochemical process that unfolds across the surface of the Earth and dissipates the abundant "free energy," or "low entropy," with which the light of the Sun floods the planet. It is made up of individuals who interact with what surrounds them, formed by structures and processes that are self-regulating, maintaining a dynamic equilibrium that persists over time. But structures and processes are not there *so that* the organisms may survive and reproduce. It is the other way around: organisms survive and reproduce *because* these structures have happened to gradually develop. They reproduce and populate the Earth *because* they are functional.

The idea goes back at least as far as Empedocles, as Darwin points out in his marvelous book.[121] Aristotle tells us in his *Physics* how Empedocles had suggested that life was the result of the casual formation of structures due to the normal combination of things. Most of these structures quickly perish, with the exception of those having the characteristics to survive: these are the living organisms.[122]

Aristotle objects that we invariably see calves born "well structured": we do not see every possible variety of shape brought forth, with only the most adequate surviving.[123] But today it has become clear that Empedocles's idea, transferred from individuals to species and enriched by what we have learned about heredity and genetics, is substantially correct.

Darwin clarified the crucial importance of the variability of biological structures that allows the continuous exploration of a space of endless possibilities; and of natural selection, which allows access to gradually more extensive regions of that space, where structures and processes are found even more capable, *together*, of persisting. Molecular biology shows us the concrete mechanism through which this happens.

The point I need to emphasize here is that having understood all this does not take away significance from notions such as "utility" and "relevance." On the contrary, it clarifies their origin, the way they are rooted in the physical world: they are the characteristics of those natural systems that *actually* give rise to survival.

These are wonderful ideas, but once again they do not explain how the notion of "meaning" emerges from the natural world. "Meaning" has intentional connotations that do not seem connected to variation and selection. The meaning of "meaning" must be based on something else.

⁂

A small miracle occurs, however, when we combine the two ideas of information and evolution.

Information plays several roles in biology. Structures and processes reproduce equal to themselves for hundreds of millions, perhaps billions of years, altered only by the slow drift of evolution. The principal means of this stability are the molecules of DNA, which remain more or less similar to their ancestors. This implies that there are *correlations*, that is to say *relative information*, across eons of time. The molecules of DNA codify and transmit information. This informational stability is perhaps the most characteristic aspect of living matter.

But there is a second way in which information is relevant in biology: in the correlations between what is inside and what is outside an organism. The majority of these correlations have no relevance for the organism. There are, however, correlations that are relevant to life in the sense in which the theory of Darwin defines relevance: favoring survival and reproduction.

I see a rock falling toward me.[124] If I move, I will survive. There is nothing mysterious about the fact that I move. It is explained by Darwin's theory: those who did not move were crushed and killed; I am a descendant of those who move out of the way. But in order to be able to move, my body needs to know that the stone is heading for me. For it to know, there must be a physical *correlation* between a physical variable inside me and the physical state of the rock. This correlation is there, obviously, because the visual system does precisely this: it correlates the surrounding environment with neural processes in the brain. There are all sorts of correlation between internal and external, but *this one* has a particular characteristic: if it was not there, or if it was not well adjusted, I would be killed by the rock. The correlation between internal and external that links the state of the rock to the neurons in my brain is directly *relevant* in the Darwinian sense: its presence or absence influences my survival.

A bacterium has a cellular membrane capable of detecting glucose gradients on which the bacterium feeds, lashes capable of making it swim, and a biochemical mechanism that points it in the direction in which there is the most glucose. The biochemistry of the membrane determines a correlation between the distribution of glucose and its internal biochemical state, which, in turn, determines the direction in which the bacterium swims. The correlation is relevant: if interrupted, the bacterium is without nourishment, and its chances of survival are diminished. There is a physical correlation with survival value.

The existence of such relevant correlations reveals the physical foundation of the notion of meaning: *relevant relative information*. Relative information in the (physical) sense given by Shannon—which is relevant in the (biological, therefore ultimately also physical) sense clarified by Darwin. This is a precise way in which we can say that its information on the concentration of sugar has *meaning* for the bacterium. Or that the thought of the tiger in my brain, that is, the corresponding neuronal configuration, actually *signifies* the tiger, an existential threat. It is correlation that matters: about which the organism "cares."[125]

Defined in this way, the notion of relevant information is physical, but also intentional in Brentano's sense. It is a connection between something (internal) and something else (generally external). It naturally carries with it a notion of "truth" or "correctness": in every particular situation, the internal state of the bacterium may encode the glucose gradient correctly, or not. It therefore has many of the ingredients that characterize "meaning."

Obviously, we also use the word "meaning" in contexts that do not have any direct relevance to survival. A poem is full of meaning, but does not seem to help my survival or reproducing probability much (or perhaps it might: a young woman might fall in love with my romantic soul . . .). The whole spectrum of what we call "meaning" in logic, psychology, linguistics, ethics and so on is not reducible to information that is *directly* relevant. But this rich spectrum has developed in the biological and cultural history of our species *starting from* something that has roots in physics, before adding the articulations/connections of our vast neural, social, linguistic and cultural complexity. This something is *relevant relative information*.

The notion of relevant relative information, in other words, is not the whole chain between physics and the full notion of meaning in the mental world, but it is the first link in this chain—the difficult one. It is the first step from the physical world, where there is nothing that corresponds to the notion of meaning, toward the world of the mind, whose grammar is based on meanings: signals that have meaning. Adding the articulations and the contexts that characterize us—the brain and its capacity to manipulate concepts (that is, processes that have meaning), our emotional states, the brain's capacity to relate to mental processes of others, our language, society, norms—we obtain something that gets gradually ever closer to the various, more complete notions of meaning.

Once we have found the first connection between physical notions and meaning, the rest follows recursively: any correlation that contributes to *directly* relevant information is also meaningful, and so on. Evolution has clearly made use of all this.

This observation clarifies why we can only speak of meaning in the context of biological processes or processes rooted in biology. But it also grounds the notion of meaning in the physical world. Meaning is not external to the natural world. We can speak of intentionality without leaving the realm of naturalism. Meaning connects something with something else, *it is a physical link* that *plays a biological role*. This is what makes an element of nature a relevant sign of something else.

And so finally I can get to the point: If we think about the physical world in terms of simple matter with variable properties, correlations are accessory facts. It seems necessary to add something extraneous to matter to speak about those correlations. But quantum physics is the discovery that the physical world is a web of correlations: relative information. The things of nature are not collections of isolated elements in haughty individualism. Meaning and intentionality are only particular cases of the ubiquity of correlations. There is a continuity between the world of meanings in our mental life and the physical world. Both are relations.

The distance between the way we think about the physical world and the way we think about our mental world diminishes.

Relative information between two objects means that if I observe the two objects, I find correlations: "You have information about the color of the sky today" means that if I ask you about the color of the sky, I find that what you tell me fits with what I see; there is a correlation between you and the sky. That two objects (the sky and you) have relative information is hence, in the final analysis, something that regards a third object (me observing you). Relative information, remember, is a dance for three, like entanglement.

But if an entity (you) is sufficiently complex to make calculations and predictions (an animal, a human being, a machine built by our technology), the fact of "having information" *also* implies having resources to make predictions. If you have information on the color of the sky, and you shut your eyes, you can *predict* what you will see when you open your eyes, even before looking: a blue sky. You have "information" on the color of the sky in a stronger sense of "information": you know beforehand what you will see.

Therefore, the elementary notion of *relative information* is the physical structure on which other, more complex notions of information are based. These now have semantic value.

Among these is the notion of information that refers to ourselves studying the rest of the physical world.

In order to be coherent, a vision of the world—a theory of the world—must be able to justify and give an account of the ways in which the inhabitants of that world arrive at that vision, at that theory.

This condition, which is perhaps a problem for naive materialism, is beautifully satisfied if we rethink matter as interaction and correlations.

My knowledge of the world is nothing other than an example of the result of interactions that generate meaningful information. It is a correlation between the external world and my memory. If the sky is blue, in my memory there is an image of a blue sky. My memory has the resources to permit me to predict the color of the sky if I close my eyes and then reopen them. Now the information I have on the sky has a *semantic* value. We know what it *means* that the sky is blue: we recognize this *meaning* when we reopen our eyes.

This is the sense of "information" I used in the postulates of quantum mechanics at the end of Chapter IV.

The double meaning of "information" gives it its ambiguous character. The basis that we have for understanding the world is our information about the world, which is effectively a (useful) correlation between us and the world. We know the world from within it.

THE WORLD SEEN FROM WITHIN

I close this chapter by mentioning one further way in which the rethinking of reality suggested by quantum theory helps us dispel the myth of a radical difference between the mental world and the physical world.

The problem of the distance between the mental and physical may seem intuitively clear, but it is difficult to delineate with precision. The mental world has different aspects—meaning, intentionality, values, objectives, ends, emotions, aesthetic and moral senses, mathematical intuition, perception, creativity, consciousness . . . Our mind does many things—it remembers, anticipates, reflects, deduces, is moved, is angered, dreams, hopes, sees; it expresses itself, imagines, creates, recognizes, knows, is self-aware . . . Taken individually, many, if not all, of these human cerebral activities do not seem so far removed from those we can more or less easily reproduce in a sufficiently complex physical device. Is there also anything that *cannot* emerge from the physics we know?

David Chalmers divided the problem of consciousness into two parts, which he called the "easy" and the "hard" problems.[126] The problem that he calls "easy" is anything but: it is how our brain functions. How, that is, it gives rise to the various behaviors that we associate with our mental life. The problem that he calls "hard" is understanding the *subjective feeling* that accompanies what the brain does.

Chalmers judges it to be plausible that the "easy" problem can be resolved in the context of our current physical conception of the world, but doubts that the same thing can be said for the "hard" one.

He asks us to imagine a "zombie," namely a machine capable of reproducing any behavior of a human being that can be observed (even with a microscope); a machine indistinguishable from a human being through any *external* observation, but which lacks subjective experience. "Inside," as Chalmers puts it, "there is no one."

The very fact that we can conceive of such possibility shows, for Chalmers, that there is a "something else" that distinguishes a living being from the zombie that could reproduce all its observable characteristics without subjective feelings. This "something else," according to Chalmers, identifies the difficulty of accounting for subjective experience in terms of our current conception of the physical world. This, for him, is the problem of consciousness.

Neuroscience is making remarkable progress in understanding the functioning of our brain. Most of its workings will probably be clarified sooner or later. Is there anything remaining that will have escaped us, after we have understood this? Chalmers maintains that there will be, because the "hard problem" is not to understand how cerebral activities work; it is to understand how these activities are accompanied by corresponding subjective feelings as they happen. That is, in order to understand the relation between our mental life and the physical world, it is essential to take into account the fact that we describe the physical world from the outside, while our mental activity is experienced in the first person, from within.

But the rethinking of the world suggested by quantum physics, it seems to me, changes the terms of the question. If the world consists of relations, then no description is from outside it. The descriptions of the world are, in the ultimate analysis, *all* from inside. They are all in the first person. Our perspective on the world, our point of view, being situated inside the world (our "situated self," as Jenann Ismael beautifully puts it[127]), is not special: it rests on the same logic on which quantum physics, hence all of physics, is based.

If we imagine the totality of things, we are imagining being *outside* the universe, looking at it from out there. But there is no "outside" to the totality of things. The external point of view is a point of view that does not exist.[128] Every description of the world is from inside it. The externally observed world does not exist; what exists are only internal perspectives on the world which are partial and reflect one another. The world *is* this reciprocal reflection of perspectives.

Quantum physics shows us that something like this happens already for inanimate things. The set of properties relative to the same object forms a perspective. If we make an abstraction from every perspective, we don't reconstruct the totality of facts; instead, we find ourselves in a world without facts, because facts are only relative facts. This is the difficulty of the Many Worlds interpretation of quantum mechanics: it describes what an external observer should expect if interacting with the world. But there are no observers external to the world. The interpretation misses the facts of the world.

Thomas Nagel, in a celebrated article, asked the question, "What is it like to be a bat?" He argued that this question is meaningful but escapes natural science.[129] The mistake, here, is to assume that physics is the description of things in the third person. On the contrary, the relational perspective shows that physics is always a first-person description of reality, from one perspective.

Ideas on the nature of the mind are often limited to just three alternatives: dualism, according to which the reality of the mind is completely different from that of inanimate things; idealism, according to which material reality only exists in the mind; and naive materialism, according to which all mental phenomena are reducible to the movement of matter. Dualism and idealism are incompatible with the discovery that we sentient beings are a part of nature like any other, and with the overwhelming and ever-increasing evidence that nothing that we observe, including ourselves, violates the natural laws that we know. Naive realism is intuitively difficult to reconcile with subjective experience.

But these are not the only alternatives. If the qualities of an object are born from the interaction with something else, then the distinction between mental and physical phenomena fades considerably. Whether it is the physical variables or what philosophers of the mind call "qualia"—elementary mental phenomena such as "I see red"—both can be thought of as more or less complex natural *phenomena*.

Subjectivity is not a qualitative leap with respect to physics: it requires a growth in complexity (Bogdanov would say of "organization"), but always in a world that is made up of perspectives, already from the most elementary level.

I think that when we wonder about the relationship between the "I" and "matter," we are using two concepts that are both confused and misleading, and this is the origin of the confusion surrounding the questions about the nature of consciousness.

Who is the "I" that has the sensation of feeling, if not the integrated set of our mental processes? We have an intuition of unity when we think about ourselves, but this is justified by the integration of our body and by the ways our mental processes work, of which the part we call conscious does one thing at a time. The first term of the problem, the "I," is the residue of a metaphysical error: the result of the common mistake of mistaking a process for an entity. (Mach is categorical: *"Das Ego ist unrettbar"*: the "I" cannot be saved. Bogdanov is said to have put it in political terms: "The individual is a bourgeois fetish."[130]) To ask what consciousness is, after having unraveled the neural processes, is like asking what a storm is after having understood its physics: it is a question that makes no sense. To add in a "possessor" of sensations is like adding Jove to the phenomenon of the thunderstorm. It is like saying, after having understood the physics of the storm, that there still remains, as Chalmers would put it, the "hard question" of connecting it with the anger of Jove.

It is true that we have the "intuition" of an independent entity that is the "I." But we also once had the "intuition" that behind a storm there was Jove. And that the Earth was flat. It is not through uncritical "intuitions" that we construct an effective comprehension of reality. Introspection is the worst instrument of inquiry if we are interested in the nature of mind: it is tantamount to looking for our own prejudices and wallowing in them.

Even worse is the second term of the question, "matter." It is, as well, the residue of an incorrect metaphysics based on too naive a conception of matter as a universal substance defined only by mass and motion. This is erroneous metaphysics because it is contradicted by quantum physics.

If we think in terms of processes, events, in terms of *relative* properties, of a world of relations, the hiatus between physical phenomena and mental phenomena is much less dramatic. It becomes possible to see both as natural phenomena generated by complex structures of interactions.

ꝉꝉ

Our knowledge of the world is articulated in various sciences, more or less connected with each other. Among the components of our knowledge, physics plays a role that quanta have partly emptied and partly enriched. On the one hand, the claim of eighteenth-century mechanism to have clarified the fundamental substance that is the basis of everything has vanished. On the other hand, the growth in our understanding of the grammar of the real has been perhaps disconcerting, but it is richer and more subtle than the previous synthesis, and it allows us to think of the world in a more effective way.

At the physical level, the world can be seen as a web of reciprocal information. In the realm of the Darwinian mechanisms, this information becomes significant, it makes sense to us. Ὁ κόσμος ἀλλοίωσις, ὁ βίος ὑπόληψις. The cosmos is change, life is discourse—as the fragment 115 by Democritus has it. The cosmos is interaction; life organizes relative information. We are a delicate and complex embroidery in the web of relations of which, as far as we currently understand it, reality is constituted.

If I look at a forest from afar, I see a dark green velvet. As I move toward it, the velvet breaks up into trunks, branches and leaves: the bark of the trunks, the moss, the insects, the teeming complexity. In every eye of every ladybug, there is an extremely elaborate structure of cells connected to neurons that guide and enable them to live. Every cell is a city, every protein a castle of atoms; in each atomic nucleus an inferno of quantum dynamics is stirring, quarks and gluons swirl, excitations of quantum fields. This is only a small wood on a small planet that revolves around a little star, among one hundred billion stars in one of the thousand billion galaxies constellated with dazzling cosmic events. In every corner of the universe we find vertiginous wells of layers of reality.

In these layers we have been able to recognize regularities and have gathered information relevant to ourselves that has enabled us to create a picture of each layer and to think about it with a certain coherence. Each one is an approximation. Reality is not divided into levels. The levels into which we break it down, the objects into which it appears to be divided, are the ways in which nature relates to us, in dynamical configurations of physical events in our brain that we call "concepts." The separation of reality into levels is relative to our way of being in interaction with it.

Fundamental physics is no exception. Nature follows its simple rules, but the complexity of things often renders the general laws irrelevant to us. Knowing that my girlfriend obeys Maxwell's equations will not help me to make her happy. When learning how a motor functions, it is best to ignore the nuclear forces between its elementary particles. There is an autonomy and independence of levels of understanding of the world that justifies the autonomy of the different areas of knowledge. In this sense, elementary physics is much less useful than physicists would like to think.

But these are not real fractures. The basics of chemistry are understandable in terms of physics, the basics of biochemistry in terms of chemistry, the basics of biology in terms of biochemistry, and so on. We understand some of these articulations well; others less so. The fractures are just gaps in our understanding. This is the sense of the question about the physical basis of the notion of meaning earlier in this chapter.

The relational perspective distances us from subject/object and matter/spirit dualisms, and from the apparent irreducibility of the reality/thought or brain/consciousness dualism. If we come to untangle the processes that take place in our bodies and their relations with the external world, what is left to understand? What is the phenomenology of our consciousness if not the name that these processes assign to themselves in the game of mirrors of relevant information contained in the signals carried by our neurons?

There remains what Chalmers calls the "easy" problem, which is anything but easy and anything but solved. We understand little about the workings of the brain. But there is no reason to suspect that in our mental life there is something not comprehensible in terms of the known natural laws.

Objections to the possibility of understanding our mental life in terms of known natural laws, on closer inspection, come down to a generic repetition of "It seems implausible to me," based on intuitions without supporting arguments.*[131] Unless it is the sad hope of being constituted by some vaporous supernatural substance that remains alive after death: a prospect that, apart from being utterly implausible, strikes me as ghastly.

As the American philosopher Erik C. Banks writes, in the quotation with which this chapter began, "How-

---

*An example of this attitude is Thomas Nagel's *Mind and Cosmos: Why the Materialist Neo-Darwinian Conception of Nature Is Almost Certainly False*, a book that obsessively repeats: "It does not seem possible to me." On a careful reading, I find that it doesn't offer any convincing argument to sustain its thesis, but rather declares ignorance, incomprehension and, especially, explicit lack of interest in the natural sciences.

ever mysterious the mind-body problem may be for us, we should always remember that it is a solved problem for nature. All we have to do is figure out that solution by naturalistic means." Quantum theory does not give us a direct solution, but it changes the terms of the question.

VII

BUT IS IT REALLY POSSIBLE?

In which I try to conclude a story that has no conclusion.

You do look, my son, in a moved sort,
As if you were dismay'd. Be cheerful, sir.
Our revels now are ended. These our actors,
As I foretold you, were all spirits and
Are melted into air, into thin air.
And, like the baseless fabric of this vision,
The cloud-capp'd towers, the gorgeous palaces,
The solemn temples, the great globe itself,
Ye all which it inherit, shall dissolve.
And, like this insubstantial pageant faded,
Leave not a rack behind. We are such stuff
As dreams are made on, and our little life
Is rounded with a sleep.

One of the most fascinating recent developments in neuroscience concerns the functioning of our visual system. How do we see? How do we know that what we have in front of us is a book, or a cat?

It would seem natural to think that receptors detect the light that reaches the retinas of our eyes and transform it into signals that race to the interior of the brain, where groups of neurons elaborate the information in ever more complex ways, until they interpret it and identify the objects in question. Neurons recognize lines that separate colors, other neurons recognize shapes drawn by these lines, others again check these shapes against data stored in our memory. Others still arrive at the recognition: it's a cat.

It turns out, however, that the brain does not work like this at all. It functions, in fact, in an opposite way. Many, if not most, of the signals do not travel from the eyes to the brain; they go the other way, from the brain to the eyes.[132]

What happens is that the brain *expects* to see something, on the basis of what it knows and has previously occurred. The brain elaborates an image of what it *predicts* the eyes should see. This information is conveyed *from* the brain *to* the eyes, through intermediate stages. If a discrepancy is revealed between what the brain expects and the light arriving into the eyes, *only then* do the neural circuits send signals toward the brain. So images from around us do not travel from the eyes to the brain—only news of discrepancies regarding what the brain expects do.

The discovery that sight functions in this way came as a surprise. But if we think about it, it becomes clear that this is the most efficient way of retrieving information from the surroundings. What would be the point of sending signals toward the brain that do nothing but confirm what it already knows? Information technology uses similar techniques to compress files of images: instead of putting into the memory the color of all the pixels, it stores information only on where the colors *change*. That is less information, but enough to reconstruct the images.

The implications for the relationship between what we see and the world, however, are remarkable. When we look around ourselves, we are not truly "observing": we are instead dreaming an image of the world based on what we know (including bias and misconception) and unconsciously scrutinizing the world to reveal any discrepancies, which, if necessary, we will try to correct.

What I see, in other words, is not a reproduction of

the external world. It is what I expect, corrected by what I can grasp. The relevant input is not that which *confirms* what we already know, but that which *contradicts* our expectations.

Sometimes it is a detail: the cat's ear has moved. Sometimes something alerts us to jump to a new hypothesis: it isn't a cat, it is a tiger! Sometimes it is a wholly new scenario, which we try to make sense of by imagining a version of it that would makes sense to us. It is in terms of what we already know that we seek to give sense to what arrives at our pupils.

This could even be a way in which the brain operates in general. In the model known as PCM (projective consciousness model), for example, the hypothesis is that consciousness is the activity of the brain constantly attempting to predict the input that constantly varies because of the variability of the world and the change of our position. Representations are techniques to minimize mistakes in predictions using observed discrepancies.[133]

In the words of the nineteenth-century French philosopher Hippolyte Taine, we can say that "external perception is an internal dream which proves to be in harmony with external things; and instead of calling 'hallucination' a false perception, we must call external perception 'a confirmed hallucination.'"[134]

Science, we may say, is only an extension of the way in which we see: we seek out discrepancies between what we expect and what we gather from the world. We have visions of the world, and if they don't work, we change them. The whole of human knowledge is constructed in this way.

Vision happens inside the brain of each of us in fractions of seconds. The growth of knowledge happens slowly, in the dense dialogue of the whole of humanity over years, decades, centuries. The first relates to the individual organization of experience and belongs to the neuronal and psychological realms. The second relates to the social organization of experience that founds the physical order described by science. (Bogdanov: "The difference between the psychological and physical orders boils down to the difference between experience organized individually and experience organized socially."[135]) But it is the same thing: we update and improve our mental maps of reality, our conceptual structure, to take into account the discrepancies we have observed between the ideas that we have and what comes to us, and hence to better and better decipher reality.[136]

Sometimes it's a detail: we learn some new fact. Sometimes we put into question the very conceptual grammar of our way of conceiving the world. We update our deepest image of the world. We discover new maps for thinking about reality that describe the world to us a little more accurately.

This is quantum theory.

ꝉꝉ

There is, of course, something bewildering about the vision of the world that emerges from this theory. We must abandon something that seemed most natural to us: the simple idea of a world made of things. We recognize it as an old prejudice, an old vehicle that we no longer have any use for.

Something of the solidity of the world seems to melt into air, like the iridescent and purplish colors of a psychedelic experience. It leaves us stunned, like the words of Prospero in the epigraph to this chapter: "And, like the baseless fabric of this vision, / The cloud-capp'd towers, the gorgeous palaces, / The solemn temples, the great globe itself, / Ye all which it inherit, shall dissolve, / And, like this insubstantial pageant faded, / Leave not a rack behind."

This is the ending of *The Tempest*, Shakespeare's last work, one of the most moving passages in the history of literature. After taking us on such a flight of imagination, taking us temporarily outside ourselves, Prospero/Shakespeare comforts us for looking "in a moved sort" and "as if you were dismay'd": "Be cheerful . . . / Our revels now are ended. These our actors, / As I foretold you, were all spirits and / Are melted into air, into thin air." Only to then softly dissolve into that immortal whisper: "We are such stuff / As dreams are made on, and our little life / Is rounded with a sleep."

This is how I feel, at the end of this long meditation on quantum physics. The solidity of the physical world seems to have melted into thin air, like Prospero's cloud-capp'd towers and gorgeous palaces. Reality has broken up into a play of mirrors.

And yet we are not talking about the sumptuous imagination of the Bard and his incursions into the hearts of humans. We are not dealing with the latest crazy speculation of some overimaginative theoretical physicist, either. It is the patient, rational, empirical, rigorous research of fundamental physics that has brought about this dissolving of substantiality. It is the best science that humanity has found to date, the basis of modern technology, whose reliability is beyond doubt.

I think it is time to take this theory fully on board, for its nature to be discussed beyond the restricted circles of theoretical physicists and philosophers, to deposit its distilled honey, sweet and intoxicating, into the whole of contemporary culture.*

I hope that what I have written may contribute to this.

The best description of reality that we have found is in terms of events that weave a web of interactions. "Entities" are nothing other than ephemeral nodes in this web. Their properties are not determined until the moment of these interactions; they exist only in relation to something else. Everything is what it is only with respect to something else.

Every vision is partial. There is no way of seeing reality that is not dependent on a perspective—no point of view that is absolute and universal.

---

*There are, of course, many lines of thought that take inspiration from or are rooted in quantum physics, more or less seriously. I find fascinating, to mention only one example, Karen Barad's utilization of the ideas of Niels Bohr in *Meeting the Universe Halfway* (Durham, NC: Duke University Press, 2007) and "Posthumanist Performativity: Toward an Understanding of How Matter Comes to Matter," *Signs: Journal of Women in Culture and Society* 28 (2003), 801–31.

And yet, points of view communicate. Knowledge is in dialogue with itself and with reality. In the dialogue, those points of view modify, enrich, converge—and our understanding of reality deepens.

The actor of this process is not a subject distinct from phenomenal reality, outside it, nor any transcendent point of view; it is a portion of that reality itself. Natural selection has taught it to make use of useful correlations: meaningful information. Our discourse on reality is itself part of that reality. *Relations* make up our "I," as our society, our cultural, spiritual and political life.

It is for this reason, I think, that everything we have been able to accomplish over the centuries has been achieved in a network of exchanges, collaborating. This is why the politics of collaboration is so much more sensible and effective than the politics of competition . . .

It is for this reason as well, I believe, that the very idea of an individual "I"—that solitary and rebellious "I" that led me to the unbridled questions of my youth, that self that I believed to be completely independent and totally free . . . recognizes itself, in the end, as only a ripple in a network of networks . . .

The questions that led me, so many years ago, to the study of physics in order to understand the structure of reality—to understand how the mind works, to understand how we comprehend reality—are still very much open. And yet we are learning. Physics has not deluded me. It has bewitched, astonished, confused and disconcerted me; given me anxious, sleepless nights looking into the dark and thinking: "But is it really possible? Can we believe this?" The question with which this book began, whispered by Časlav on the shore of the island of Lamma.

Physics seemed to me the place where the weave between the structure of reality and the structure of thought was closest, the place where this intertwinement was subject to the incandescent test of continuous evolution. The journey undertaken has been more surprising and more of an adventure than I expected. Space, time, matter, thought, the whole of reality has redesigned itself before my eyes, as in some vast kaleidoscope. Quantum physics, more than the immensity of the universe and the discovery of its great history, more even than the extraordinary vision of Einstein, has been for me the heart of this radical questioning of our mental maps of reality.

The classical vision of the world, to adapt Taine's phrase, is no longer a confirmed hallucination. The fragmentary and insubstantial quantum world is, for the moment, the hallucination most in harmony with reality . . .

There is a sense of the vertiginous—of freedom, happiness, lightness—in the vision of the world that we are offered by the discovery of quanta. "You do look . . . in a moved sort, / As if you were dismay'd: be cheerful . . ." In the end, the youthful curiosity that drew me toward physics, like a child following a magic flute, has led me to finding more enchanted castles than I could have dreamed of. The world of quantum theory that I have attempted to describe, opened by a young man's journey to the Sacred Island in the North Sea, seems extraordinarily beautiful to me.

Of Helgoland—that extreme, wind-battered place—Goethe had written that it was one of those places on Earth that "exemplifies the endless fascination of Nature." And that on the Sacred Island it was possible to experience the "world spirit," the *Weltgeist*.[137] Who knows, perhaps it was this spirit that spoke to Heisenberg, helping him clear a little of the fog from our eyes . . .

Every time that something solid is put into doubt or dismantled, something else opens up and allows us to see further than we could before. Watching what appeared to be as solid as rock melt into air makes lighter, it seems to me, the transitory and bittersweet flowing of our lives.

The interconnectedness of things, the reflection of one in another, shines with a clear light that the coldness of eighteenth-century mechanism could not capture.

Even if it leaves us astonished.
Even if it leaves us with a profound sense of mystery.

Acknowledgments

Thanks to Blu. To Emanuela, Lee, Časlav, Jenann, Ted, David, Roberto, Simon, Eugenio, Aurélien, Massimo, Enrico, for a thousand things. To Andrea for his valuable comments on a first draft of the book; to Maddalena for making these lines readable; to Sami, with nostalgia, for his support and his friendship; to Guido for having shown me the path of my life; to Bill for having been the first to listen to these things fifteen years ago; to Wayne for his insight; to Chris for his hospitality; to Antonino for the wonderful suggestions. To my father for teaching me what it means to still be there when you are no longer there. To Simone and Alejandro for having made, together, the most beautiful research group in the world. To my fantastic students, to my colleagues in physics and philosophy with whom I have

discussed the issues in this book over the years, to my marvelous readers. To all of these people, who together weave the magic net of relations of which this book is a thread.

Thanks, above all, to Werner and Aleksandr.

Notes

I. A STRANGELY BEAUTIFUL INTERIOR

The Absurd Idea of the Young Heisenberg: Observables

1. This and subsequent quotations from Heisenberg are taken, with minimal adaptation, from Werner Heisenberg, *Der Teil und das Ganze* (Munich: Piper, 1969).
2. Niels Bohr, "The Genesis of Quantum Mechanics," in *Essays on Atomic Physics and Human Knowledge 1958–1962* (New York: Wiley, 1963), 74–78.
3. Werner Heisenberg, "Über quantentheoretische Umdeutung kinematischer und mechanischer Beziehungen," *Zeitschrift für Physik* 33 (1925), 879–93.
4. Max Born and Pascual Jordan, "Zur Quantenmechanik," *Zeitschrift für Physik* 34 (1925), 858–88.
5. Paul Dirac, "The Fundamental Equations of Quantum Mechanics," *Proceedings of the Royal Society A* 109, no. 752 (1925), 642–53.
6. He realizes that Heisenberg's tables are noncommutative, which makes him think of the Poisson brackets that he had encountered in an advanced mechanics course. A delightful account of these fateful years, in the voice of a seventy-three-year-old Dirac, can be found at https://www.youtube.com/watch?v=vwYs8tTLZ24.
7. Max Born, *My Life: Recollections of a Nobel Laureate* (London: Taylor and Francis, 1978), 218.

8. Wolfgang Pauli, "Über das Wasserstoffspektrum vom Standpunkt der neuen Quantenmechanik," *Zeitschrift für Physik* 36 (1926), 336–63, a triumph of virtuoso technique.
9. Cited in F. Laudisa, *La realtà al tempo dei quanti: Einstein, Bohr e la nuova immagine del mondo* (Turin: Bollati Boringhieri, 2019), 115.
10. Albert Einstein, *Corrispondenza con Michele Besso (1903–1955)* (Naples: Guida, 1995), 242.
11. Bohr, "The Genesis of Quantum Mechanics," 75.
12. In Dirac's terms: q-numbers. In more modern terms: operators. More generally: variables of the noncommutative algebra defined by the equation discussed in the next chapter.

The Misleading ψ of Erwin Schrödinger: Probability

13. W. J. Moore, *Schrödinger, Life and Thought* (Cambridge, UK: Cambridge University Press, 1989), 131.
14. Erwin Schrödinger, "Quantisierung als Eigenwertproblem (Zweite Mitteilung)," *Annalen der Physik* 384, no. 4 (1926), 489–527.
15. That is to say, inverting the eikonal approximation.
16. Erwin Schrödinger, "Quantisierung als Eigenwertproblem (Erste Mitteilung)," *Annalen der Physik* 384, no. 4 (1926), 361–76. At first he had written the equation relativistically and was convinced that it was wrong. Then he contented himself with studying the nonrelativistic limit, and this worked.
17. Erwin Schrödinger, "Über das Verhältnis der Heisenberg-Born-Jordanschen Quantenmechanik zu der meinem," *Annalen der Physik* 384, no. 5 (1926), 734–56.
18. Throughout the book I call ψ the "wave function," that is, the quantum state in the base position is the abstract quantum state, represented by a vector in a Hilbert space. For the considerations that follow, the distinction is not relevant.
19. George Uhlenbeck, quoted in A. Pais, "Max Born's Statistical Interpretation of Quantum Mechanics," *Science* 218 (1982), 1193–98.
20. Cited in Manjit Kumar, *Quantum: Einstein, Bohr and the Great Debate about the Nature of Reality* (London: Icon Books, 2008), 155.
21. Kumar, *Quantum*, 220.

22. Erwin Schrödinger, *Nature and the Greeks and Science and Humanism* (Cambridge, UK: Cambridge University Press, 1996).
23. Max Born, "Quantenmechanik der Stossvorgänge," *Zeitschrift für Physik* 38 (1926), 803–27.
24. The squared modulus of ψ (x) gives the density of probability that the particle will be observed at point x rather than anywhere else.
25. They have changed the rules, and it is now illegal.
26. In the same way, Heisenberg's theory gives the probability that we will see something, given the previous observations.

The Granularity of the World: Quanta

27. $B=2hv^3c^{-2} / (e^{hv/kT}-1)$. Max Planck, "Über eine Verbesserung der Wienschen Spectraleichung," *Verhandlungen der Deutschen Physikalischen Gesellschaft* 2 (1900), 202–4.
28. $E = hv$.
29. Albert Einstein, "Über einen die Erzeugung und Verwandlung des Lichtes betreffenden heuristischen Gesichtspunkt," *Annalen der Physik* 322, no. 6 (1905), 132–48.
30. This is the effect on which photoelectric cells are based: on certain metals the light produces a small electric current. The strange thing is that this does not happen for light of low frequency, independently of the intensity of the light. Einstein understood that the reason is that—regardless of how many there are—the photons of lower frequency are less energetic and do not have sufficient energy to extract electrons from atoms.
31. Niels Bohr, "On the Constitution of Atoms and Molecules," *Philosophical Magazine and Journal of Science* 26 (1913), 1–25.
32. Subsequently published in Niels Bohr, "The Quantum Postulate and the Recent Development of Atomic Theory," *Nature* 121 (1928), 580–90.
33. P. A. M. Dirac, *The Principles of Quantum Mechanics* (Oxford: Oxford University Press, 1930).
34. J. von Neumann, *Mathematische Grundlagen der Quantenmechanik* (Berlin: Springer, 1932).
35. J. Bernstein, "Max Born and the Quantum Theory," *American Journal of Physics* 73 (2005), 999–1008.

II. A CURIOUS BESTIARY OF EXTREME IDEAS

Superpositions

36. P. A. M. Dirac, *I principi della meccanica quantistica* (Turin: Bollati Boringhieri, 1968); L. D. Landau and E. M. Lisfits, *Meccanica quantistica* (Rome: Editori Riuniti, 1976); R. Feynman, *La fisica di Feynman: The Feynman Lectures on Physics*, vol. 3 (London: Addison-Wesley, 1970); *La fisica di Berkeley*, vol. 4, *Fisica quantistica* (Bologna: Zanichelli, 1973); A. Messiah, *Quantum Mechanics*, vol. 1 (Amsterdam: North Holland, 1967).
37. Cited in A. Pais, *Ritratti di scienziati geniali. I fisici del XX secolo* (Turin: Bollati Boringhieri, 2007), 31.
38. Erwin Schrödinger, "Die gegenwärtige Situation in der Quantenmechnik," *Naturwissenschaften* 23 (1935), 807–12.
39. This is why we do not become aware of quantum mechanics in our daily lives. We do not see the effects of interference and therefore can replace the quantum superposition between cat-awake and cat-asleep with the simple fact that we do not know whether the cat is asleep or not. The suppression of interference phenomena for objects that interact with a large number of variables is well understood. It is called "quantum decoherence."

Taking ψ Seriously:
Many Worlds, Hidden Variables and Physical Collapses

40. Many books reconstruct this historic discussion in more detail. For instance, see the excellent *Quantum* by Manjit Kumar and more recently *La realtà al tempo dei quanti* by Federico Laudisa. Laudisa is sympathetic to Einstein's intuition; I follow more in the footsteps of Bohr and Heisenberg.
41. David Kaiser, *How the Hippies Saved Physics: Science, Counterculture, and the Quantum Revival* (New York: W. W. Norton, 2012).
42. For a recent defense of this interpretation, see Sean Carroll, *Something Deeply Hidden: Quantum Worlds and the Emergence of Spacetime* (New York: Dutton, 2019).

43. It is not enough to know the ψ wave and Schrödinger's equation in order to define and use quantum theory: we need to specify an algebra of observables, otherwise we cannot calculate anything and there is no relation with the phenomena of our experience. The role of this algebra of observables, which is extremely clear in other interpretations, is not at all clear in the Many Worlds interpretation.
44. A presentation and a defense of Bohm's theory can be found in David Z. Albert, *Quantum Mechanics and Experience* (Cambridge, MA, and London: Harvard University Press, 1992).
45. The way in which we interact with the particle is quite subtle, and often not very clear in presentations of the theory. The wave of an instrument of measurement interacts with the wave of the electron, but the dynamic of the equipment is guided by the value of the common wave determined by the position of the electron, and hence its evolution is determined by where the electron actually is.
46. There is another possibility: that quantum mechanics is only an approximation, and the hidden variables effectively reveal themselves in some other regime. For now, these modifications of the predictions of quantum mechanics, however, cannot be seen.
47. The space of the configurations of the set of particles.
48. There are different versions of these theories, all somewhat artificial and incomplete. There are two that are better known: a concrete mechanism designed by the Italian physicists Giancarlo Ghirardi, Alberto Rimini and Tullio Weber; and Roger Penrose's hypothesis that the collapse is induced by gravity when the quantum superposition between different configurations of space-time exceeds a threshold value.

Accepting Indeterminacy

49. C. Calosi and C. Mariani, "Quantum Relational Indeterminacy," *Studies in History and Philosophy of Science. Part B: Studies in History and Philosophy of Modern Physics* 71 (2020), 158–69.
50. More precisely, the quantity ψ is like the Hamilton's function S (the solution of the Hamilton–Jacobi equation) in classical mechanics: a

calculation tool, not an entity to be considered real. As evidence of this, observe that Hamilton's *S* function is effectively the classical limit of the wave function: $\psi \sim \exp iS/\hbar$.

51. In the sense of Johann Gottlieb Fichte, Friedrich Schelling and G. W. F. Hegel.

III. IS IT POSSIBLE THAT SOMETHING IS REAL IN RELATION TO YOU BUT NOT IN RELATION TO ME?

Relations

52. For a technical introduction to the relational interpretation of quantum mechanics, see the entry "Relational Quantum Mechanics" in *Stanford Encyclopedia of Philosophy*, ed. E. N. Zalta, https://plato.stanford.edu/archives/win2019/entries/qm-relational/.
53. Niels Bohr, *The Philosophical Writings of Niels Bohr* (Woodbridge, CT: Ox Bow Press, 1998), vol. 4, 111.
54. The properties I am referring to are those that are variables: that is, those described by functions on the phase space, not the invariant properties such as the nonrelativistic mass of a particle.
55. An event is real with respect to a stone if it acts on it, if it modifies it. An event is not real with respect to the stone if its fall implies that interference phenomena that occur do not occur.

The Rarefied and Subtle World of Quanta

56. An event *e1* is "relative to A, but not to B," in the following sense: *e1* acts on A, but there is an event *e2* that can act on B which would be impossible if *e1* had acted on B.
57. The first person to realize the relational character of the ψ wave was a young American doctoral student in the mid-1950s, Hugh Everett III. His doctoral thesis, "The Formulation of Quantum Mechanics Based on Relative States," has a had a big influence on discussions of quanta.
58. Anthony Aguirre, *Cosmological Koans: A Journey to the Heart of Physical Reality* (New York: W. W. Norton, 2019), chap. 44.

59. Erwin Schrödinger, *Nature and the Greeks and Science and Humanism* (Cambridge, UK: Cambridge University Press, 1996).
60. Carlo Rovelli, *Che cos'è la scienza. La rivoluzione di Anassimandro* (Milan: Mondadori, 2011).

IV. THE WEB OF RELATIONS THAT WEAVES REALITY

Entanglement

61. Juan Yin et al., "Satellite-Based Entanglement Distribution over 1200 Kilometers," *Science* 356 (2017), 1140–44.
62. John S. Bell, "On the Einstein–Podolsky–Rosen Paradox," *Physics Physique Fizika* 1 (1964), 195–200.
63. Bell's argument is subtle, very technical, but solid. An interested reader can find it, with extensive detail, in *Stanford Encyclopedia of Philosophy*: https://plato.stanford.edu/entries/bell-theorem/.
64. If ψ_1 is the Schrödinger wave for an object and ψ_2 is the wave of a second object, our intuition tells us that we only need to know ψ_1 and ψ_2 in order to predict everything that it is possible to observe about the two objects. But this is not the case. Schrödinger's wave for two objects is not the same as the two individual waves. It is a more complicated one that contains other information. The information on possible quantum correlations cannot be written with the two ψ_1 and ψ_2 waves alone. Formally, the state of two systems does not live in the tensor sum of two Hilbert $H_1 \oplus H_2$, but rather in their tensor product $H_1 \otimes H_2$. The general form of the wave function of two systems in any base is not $\psi_{12}(x1,x2) = \psi_1(x1)\psi_2(x2)$ but a generic function $\psi_{12}(x1,x2)$, which can be a quantum superposition of terms of the form $\psi_{12}(x1,x2) = \psi_1(x1)\psi_2(x2)$; that is, it includes entangled states.
65. In the language of analytic philosophy, the relation does not supervene on the state of the single objects. It is necessarily external, not internal.

The Dance for Three That Weaves the Relations of the World

66. The reason is that in the entangled state $|A\rangle \otimes |OA\rangle + |B\rangle \otimes OB\rangle$ where A and B are the observed properties and OA and OB are the

observer's variables correlated with these, a measurement of A collapses the system to the state $|A\rangle \otimes |OA\rangle$, and therefore a later measure of the observer's variables yields OA.

67. The tensorial structure of the Hilbert spaces of the subsystems.

Information

68. This is the definition of "relative information" given by Claude Shannon in the classic work that introduced the theory of information: Claude E. Shannon, "A Mathematical Theory of Communication," *Bell System Technical Journal* 27 (1948), 379–423. Shannon insisted that his definition has nothing mental or semantic about it.
69. These postulates were introduced in Carlo Rovelli, "Relational Quantum Mechanics," *International Journal of Theoretical Physics* 35 (1996), 1637–78, https://arxiv.org/abs/quant-ph/9609002.
70. The phase space of which has finite Liouville volume. Every physical system can be approximated appropriately with a phase space of finite volume.
71. For example, if you measure the spin of a particle of ½ spin along two different directions, the result of the second measurement renders the result of the first irrelevant for predicting the results of future measurements of spin.
72. Ideas similar to those introduced in Rovelli, "Relational Quantum Mechanics," have appeared independently in Anton Zeilinger, "On the Interpretation and Philosophical Foundation of Quantum Mechanics," *Vastakohtien todellisuus: Festschrift for K. V. Laurikainen*, ed. U. Ketvel et al. (Helsinki: Helsinki University Press, 1996); Časlav Brukner and Anton Zeilinger, "Operationally Invariant Information in Quantum Measurements," *Physical Review Letters* 83 (1999), 3354–57.
73. More precisely: no degree of freedom of any physical system can have its state localized in its phase space with precision greater than $\hbar$ (the constant $\hbar$ has the dimensions of a volume in phase space).
74. Werner Heisenberg, "Über den anschaulichen Inhalt der quantentheoretischen Kinematik und Mechanik," *Zeitschrift für Physik* 43 (1927), 172–98.

75. At first Heisenberg and Bohr interpreted in a concrete way the fact that to measure one variable altered another: because of granularity, no measurement—they thought—could be sufficiently delicate to not modify the object observed. But Einstein, with insistent criticism, forced them to recognize that things were more subtle. Heisenberg's principle does not mean that position and velocity have definite values and that we cannot know both because to measure one modifies the other. It means that a quantum particle is something that never has perfectly determined position and velocity. They are determined only in an interaction, at the price of rendering one or the other indeterminate.
76. The observables form a noncommutative algebra.
77. This fact is clarified well by the phenomenon of "quantum decoherence," whereby quantum interference phenomena are not seen in an environment with many variables.
78. This point is clarified in the paper by Andrea Di Biagio and Carlo Rovelli titled "Relative Facts, Stable Facts," https://arxiv.org/abs/2006.15543.
79. The central limit theorem. In its simplest version, it states that the fluctuation of the sum of N variables commonly grows as $\sqrt{N}$, and this implies a mean fluctuation in the order of $\sqrt{N}/N$ which goes to zero for large N.

V. THE UNAMBIGUOUS DESCRIPTION OF AN OBJECT INCLUDES THE OBJECTS TO WHICH IT MANIFESTS ITSELF

Aleksandr Bogdanov and Vladimir Lenin

80. V. Il'in, *Materializm i empiriokriticizm* (Moscow: Zveno, 1909); translated as V. Lenin, *Materialism and Empirio Criticism: Collected Works of V. I. Lenin*, vol. 13 (Whitefish, MT: Literary Licensing, 2011).
81. See for instance (although I disagree with some of its conclusions) David Bakhurst, "On Lenin's Materialism and Empiriocriticism," *Studies in East European Thought* 70 (2018), 107–19, https://doi.org/10.1007/s11212-018-9303-7, and references therein.
82. A. Bogdanov, *Empiriomonizm: Stat'i po filosofi* (Moscow and St. Petersburg: S. Dorovatovskij and A. Čarušnikov, 1904–06); trans-

lated by David Rowley as *Empiriomonism: Essays in Philosophy, Books 1–3* (Leiden: Brill, 2019).

83. An acute summary of Mach's ideas and an interesting reevaluation of his thought can be found in Erik C. Banks, *The Realistic Empiricism of Mach, James and Russell: Neutral Monism Reconceived* (Cambridge, UK: Cambridge University Press, 2014).
84. "A barometric low hung over the Atlantic. It moved eastward toward a high-pressure area over Russia without as yet showing any inclination to bypass this high in a northerly direction. The isotherms and isotheres were functioning as they should. The air temperature was appropriate relative to the annual mean temperature and to the aperiodic monthly fluctuations of the temperature. The rising and setting of the sun, the moon, the phases of the moon, of Venus, of the rings of Saturn, and many other significant phenomena were all in accordance with the forecasts in the astronomical yearbooks. The water vapor in the air was at its maximal state of tension, while the humidity was minimal. In a word that characterizes the facts fairly accurately, even if it was a bit old-fashioned: It was a fine day in August 1913." Robert Musil, *The Man without Qualities* (1930–43), trans. Sophie Wilkins (London: Picador, 1995).
85. Friedrich Adler, *Ernst Machs Überwindung des mechanischen Materialismus* (Vienna: Brand, 1918).
86. Ernst Mach, *Die Mechanik in ihrer Entwicklung historischkritisch dargestellt* (Leipzig: Brockhaus, 1883).
87. Banks, *The Realistic Empiricism of Mach, James and Russell.*
88. Bertrand Russell, *The Analysis of Mind* (London and New York: Allen & Unwin/Macmillan, 1921), 10.
89. Aleksandr Bogdanov, "Vera i nauka O knige V. Il'ina *Materializm i empiriokriticizm*," in *Padenie velikogo fetišma (Sovremnnyj krizis ideologi)* [The Fall of a Great Fetishism (The Contemporary Ideological Crisis)], (Moscow: S. Dorovatovskij and A. Čarušnikov, 1910). A detailed discussion of Mach's ideas is found in Aleksandr Bogdanov, *Priključenija odnoj filosofskoj školy* (St. Petersburg: Znanie, 1908). Works of Bogdanov translated into English can be found at https://www.marxists.org/archive/bogdanov/index.htm. See a full bibliography at https://monoskop.org/Alexander_Bogdanov#Links.

90. Popper also badly misreads Mach along similar lines: Karl Popper, "A Note on Berkeley as Precursor of Mach and Einstein," *British Journal for the Philosophy of Science* 4 (1953), 26–36.
91. "The only property of matter to which the philosophical position of materialism is linked is that of being an objective reality, of existing outside of our minds." Lenin, *Materialism and Empirio Criticism*, chap. 5.
92. Ernst Mach, *The Science of Mechanics*, trans. Thomas J. McCormack (La Salle, IL: Open Court, 1960), 559.
93. And if that is not enough, reread the footnote to paragraph 4.9 of Mach, *The Science of Mechanics* (589–90): it seems like a diligent explanation by a good student of the idea that forms the basis of Einstein's general relativity. Except that it was written in 1883 . . . thirty-two years before Einstein published his theory.
94. Bertram D. Wolfe, *Three Who Made a Revolution: A Biographical History of Lenin, Trotsky and Stalin* (Boston: Beacon Press, 1962), 517; David Bakhurst, "On Lenin's Materialism and Empiriocriticism." *Studies in East European Thought* 70 (2018), 107–19.
95. Douglas W. Huestis, "The Life and Death of Alexander Bogdanov, Physician," *Journal of Medical Biography* 4 (1996), 141–47.
96. Alexander Aleksandrovich Bogdanov, "Bogdanov's Autobiography," from *Empiriomonism: Essays in Philosophy, Books 1–3*, trans. David Rowley (Leiden: Brill, 2019), https://brill.com/view/book/edcoll/9789004300323/front-7.xml.
97. Bakhurst, "On Lenin's Materialism and Empiriocriticism."
98. Wu Ming, *Proletkult* (Turin: Einaudi, 2018); Kim Stanley Robinson, *Red Mars, Green Mars, Blue Mars* (New York: Spectra, 1993–96).

Naturalism without Substance: Contextuality

99. Douglas Adams, speech given at Digital Biota 2, Cambridge, UK, September 1998, http://www.biota.org/people/douglasadams/index.html. Found also in *The Salmon of Doubt: Hitchhiking the Galaxy One Last Time* (New York: Del Rey, 2005).
100. For example, his response to Einstein's objection when presented with the ideal experiment of the light box is wrong. Bohr invokes

general relativity, but this has nothing to do with the question, which is about an entanglement between two distant objects.

101. Niels Bohr, *The Philosophical Writings of Niels Bohr* (Woodbridge, CT: Ox Bow Press, 1998), vol. 4, 11.
102. Mauro Dorato, "Bohr Meets Rovelli: A Dispositionalist Account of the Quantum Limits of Knowledge," *Quantum Studies: Mathematics and Foundations* 7 (2020), 133–45, https:// doi.org/10.1007/s40509-020-00220-y.
103. For Aristotle the relation is a property of the substance. It is the property of substance that refers to something else (*Categories*, 7, 6a36–7). Among all the categories, for Aristotle, relationality is the one that has "least being and reality" (*Metaphysics*, 14, 1, 1088a22–4 and 30–35). Can we think differently?

Without Foundation? Nāgārjuna

104. Carlo Rovelli, "Relational Quantum Mechanics," *International Journal of Theoretical Physics* 35 (1996), 1637–78, https://arxiv.org/abs/quant-ph/9609002; and the entry "Relational Quantum Mechanics" in *Stanford Encyclopedia of Philosophy*, ed. E. N. Zalta, at plato.stanford.edu/archives/win2019/entries/qm-relational/.
105. Bas C. van Fraassen, "Rovelli's World," *Foundations of Physics* 40 (2010), 390–417, https://www.princeton.edu/~fraassen/abstract/Rovelli_sWorld-FIN.pdf.
106. Michel Bitbol, *De l'intérieur du monde: Pour une philosophie et une science des relations* (Paris: Flammarion, 2010). Relational quantum mechanics is discussed in the second chapter.
107. François-Igor Pris, "Carlo Rovelli's Quantum Mechanics and Contextual Realism," *Bulletin of Chelyabinsk State University* 8 (2019), 102–7; Pierre Livet, "Processus et connexion," in *Le renouveau de la métaphysique*, ed. S. Berlioz, F. Drapeau Contim and F. Loth (Paris: Vrin, 2020).
108. Mauro Dorato, "Rovelli's Relational Quantum Mechanics, Anti-Monism, and Quantum Becoming," in *The Metaphysics of Relations*, ed. A. Marmodoro and D. Yates (Oxford: Oxford University Press, 2016), 235–62, http://arxiv.org/abs/1309.0132. On reality made up

of structures, see, for example, Steven French and James Ladyman, "Remodeling Structural Realism: Quantum Physics and the Metaphysics of Structure," *Synthese* 136 (2003), 31–56; Steven French, *The Structure of the World: Metaphysics and Representation* (Oxford: Oxford University Press, 2014).

109. Laura Candiotto, "The Reality of Relations," *Giornale di Metafisica* 2 (2017), 537–51, philsci-archive.pitt.edu/14165/.

110. Mauro Dorato, "Bohr Meets Rovelli: A Dispositionalist Account of the Quantum Limits of Knowledge," *Quantum Studies: Mathematics and Foundations* 7 (2020), 133–45, https:// doi.org/10.1007/s40509-020-00220-y.

111. Juan J. Colomina-Almiñana, *Formal Approach to the Metaphysics of Perspectives: Points of View as Access* (Heidelberg: Springer, 2018); Antti E. Hautamäki, *Viewpoint Relativism: A New Approach to Epistemological Relativism Based on the Concept of Points of View* (Berlin: Springer, 2020).

112. On structural realism, see: Steven French and James Ladyman, "In Defence of Ontic Structural Realism," in *Scientific Structuralism*, ed. A. Bokulich and P. Bokulich (Dordrecht: Springer, 2011), 25–42; James Ladyman et al., *Everything Must Go: Metaphysics Naturalized* (Oxford: Oxford University Press, 2007). Also see James Ladyman, "The Foundations of Structuralism and the Metaphysics of Relations," in *The Metaphysics of Relations*, ed. A. Marmodoro and D. Yates (Oxford: Oxford University Press, 2016), 177–97.

113. Bitbol, *De l'intérieur du monde*.

114. Laura Candiotto and Giacomo Pezzano, *Filosofia delle relazioni* (Genoa: Il Nuovo Melangolo, 2019).

115. Plato, *Sophist*, 247d–e.

116. Carlo Rovelli, *The Order of Time*, trans. Erica Segre and Simon Carnell (London: Allen Lane, 2017).

117. Erik C. Banks, *The Realistic Empiricism of Mach, James and Russell: Neutral Monism Reconceived* (Cambridge, UK: Cambridge University Press, 2014).

118. Nāgārjuna, *Mūlamadhyamakakārikā*, trans. J. L. Garfield, *The Fundamental Wisdom of the Middle Way: Nāgārjuna's "Mūlamadhyamakakārikā"* (Oxford: Oxford University Press, 1995).

119. Nāgārjuna, *Mūlamadhyamakakārikā*, XVIII, 7.

VI. "FOR NATURE IT IS A PROBLEM ALREADY SOLVED"

Simple Matter?

120. Banks, *The Realistic Empiricism of Mach, James and Russell*, chap. 5.

What Does "Meaning" Mean?

121. Charles Darwin, *The Origin of Species by Means of Natural Selection* (London: John Murray, 1859).
122. "[There could be] beings for which what happens seems organized for a purpose, when in reality things have been randomly structured and those things that were not adequately organized have perished, as Empedocles says." Aristotle, *Physics*, II, 8, 198b29.
123. Aristotle, *Physics*, II, 8, 198b35.
124. This chapter follows closely my technical article "Meaning and Intentionality = Information + Evolution," in Anthony Aguirre, *Wandering towards a Goal*, ed. B. Foster and Z. Merali (Cham, Switzerland: Springer, 2018), 17–27. The example and the idea were inspired by a lecture by David Wolpert titled "Observers as Systems That Acquire Information to Stay Out of Equilibrium," delivered at the conference "The Physics of the Observer" in Banff, Canada, in 2016.
125. Here in a sense close to the one in which it is used in Martin Heidegger, *Sein und Zeit*, in Heidegger's *Gesamtausgabe*, ed. F.-W. von Herrmann, vol. 2 (Frankfurt: Vittorio Klostermann, 1977).

The World Seen from Within

126. David J. Chalmers, "Facing Up to the Problem of Consciousness," *Journal of Consciousness Studies* 2 (1995), 200–219.
127. Jenann T. Ismael, *The Situated Self* (Oxford: Oxford University Press), 2007.
128. Mauro Dorato, "Rovelli's Relational Quantum Mechanics, Anti-Monism, and Quantum Becoming," in *The Metaphysics of Relations*, ed. A. Marmodoro and D. Yates (Oxford: Oxford University Press, 2016), 235–62, http://arxiv.org/abs/1309.0132.

129. Thomas Nagel, "What Is It Like to Be a Bat?," *Philosophical Review* 83 (1974), 435–50.
130. David Bakhurst, "On Lenin's Materialism and Empiriocriticism," *Studies in East European Thought* 70 (2018), 107–19, https://doi.org/10.1007/s11212-018-9303-7.
131. Thomas Nagel, *Mind and Cosmos: Why the Materialist Neo-Darwinian Conception of Nature Is Almost Certainly False* (Oxford: Oxford University Press, 2012).

VII. BUT IS IT REALLY POSSIBLE?

132. See, for example, Andy Clark, "Whatever Next? Predictive Brains, Situated Agents, and the Future of Cognitive Science," *Behavioral and Brain Sciences* 36 (2013), 181–204.
133. David Rudrauf et al., "A Mathematical Model of Embodied Consciousness," *Journal of Theoretical Biology* 428 (2017), 106–31; Kenneth Williford, Daniel Bennequin, Karl Friston and David Rudrauf, "The Projective Consciousness Model and Phenomenal Selfhood," *Frontiers in Psychology* 9, article 2571 (2018).
134. Hippolyte Taine, *De l'intelligence* (Paris: Librairie Hachette, 1870), 13.
135. A. Bogdanov, *Empiriomonizm: Stat'i po filosofi* (Moscow and St. Petersburg: S. Dorovatovskij and A. Čarušnikov, 1904–06), 28; translated by David Rowley as *Empiriomonism: Essays in Philosophy, Books 1–3* (Leiden: Brill, 2019).
136. The relationship between vision and science is developed in the lecture on "Appearance and Physical Reality," https://lectures.dar.cam.ac.uk/video/100/appearance-and-physical-reality, forthcoming in the collection of Darwin College lectures, *Vision* (Cambridge, UK: Cambridge University Press).
137. Johann W. von Goethe to Kaspar von Sternberg, January 4, 1831, and to Carl Friedrich Zelter, October 24, 1827, in *Gedenkausgabe der Werke, Briefe und Gespräche*, ed. E. Beutler (Zurich: Artemis, 1951), vol. 21, 767, 958.

Illustration Credits

Pages 19, 48: diagram images © Snap2Art/Shutterstock.com

Pages 54, 61: © robodread/stock.adobe.com, ElenaShow/Shutterstock.com

Pages 58, 68: diagram images © ElenaShow/Shutterstock.com, Abstract man 24/Shutterstock.com

Pages 65, 203: © robodread/stock.adobe.com

Page 81: diagram images © Anton Belo/Shutterstock.com, ElenaShow/Shutterstock.com, Abstract man 24/Shutterstock.com

Page 84: diagram images © Clipart.Email, perapong/stock.adobe.com, Anton Belo/Shutterstock.com, Serz_72/Shutterstock.com, ElenaShow/Shutterstock.com, Abstract man 24/Shutterstock.com

Page 206, left: Werner Heisenberg, 1924 © 2020 Foto Scala, Firenze/bpk, Bildagentur für Kunst, Kultur und Geschichte, Berlin

Index

A NOTE ABOUT THE TYPE

The text of this book was set in Garamond Premier Pro. Garamond Premier Pro was designed by Robert Slimbach on the model of the roman types of Claude Garamond and the italic types of Robert Granjon; it represents a reworking and expansion of the earlier Garamond Pro.

The display of this book was set in Neutraface 2. Neutraface is a geometric sans-serif typeface designed by Christian Schwartz for House Industries, an American digital type foundry. It was influenced by the work of architect Richard Neutra and was developed with the assistance of Neutra's son and former partner, Dion Neutra.

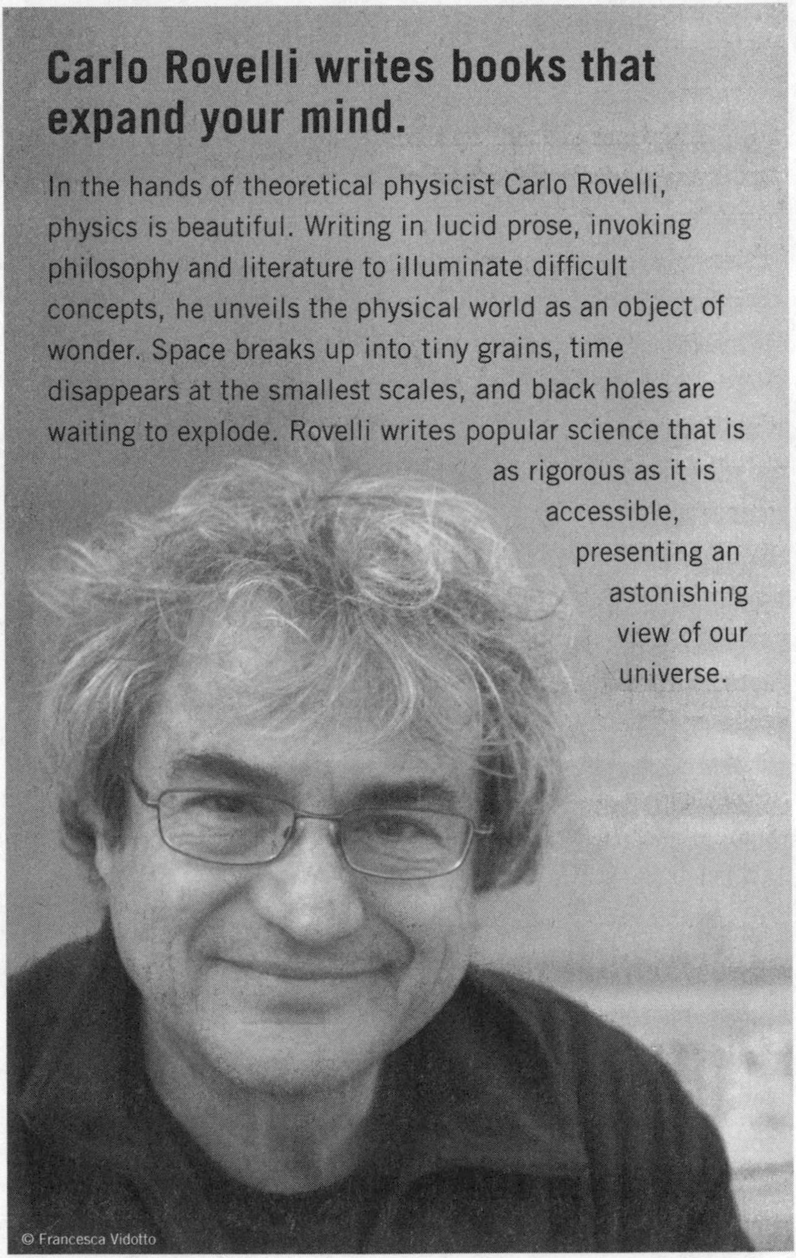
Carlo Rovelli writes books that expand your mind.
In the hands of theoretical physicist Carlo Rovelli, physics is beautiful. Writing in lucid prose, invoking philosophy and literature to illuminate difficult concepts, he unveils the physical world as an object of wonder. Space breaks up into tiny grains, time disappears at the smallest scales, and black holes are waiting to explode. Rovelli writes popular science that is as rigorous as it is accessible, presenting an astonishing view of our universe.
© Francesca Vidotto

A closer look at the mind-bending nature of the universe

What are the elementary ingredients of the world? Do time and space exist? And what exactly is reality? Carlo Rovelli leads us on a wondrous journey from Democritus to Einstein, from Michael Faraday to gravitational waves, and from classical physics to his own work in quantum gravity. As he shows us how the idea of reality has evolved over time, Rovelli offers deeper explanations of the theories he introduced so concisely in *Seven Brief Lessons on Physics.*

"The man who makes physics sexy . . . the scientist they're calling the next Stephen Hawking." ***—The Times Magazine* (UK)**

"Rovelli's book is a gem. It's a pleasure to read, full of wonderful analogies and imagery and, last but not least, a celebration of the human spirit." **—NPR's *Cosmos & Culture***

A concise, elegant exploration of time that weaves the results of contemporary physics together with insights from philosophy and literature

Why do we remember the past and not the future? Do we exist in time, or does time exist in us? In accessible prose, Carlo Rovelli tears down our assumptions about time one by one, revealing a strange universe where, at the most fundamental level, time disappears. *The Order of Time* offers a profoundly intelligent, culturally rich new appreciation of the mysteries of time.

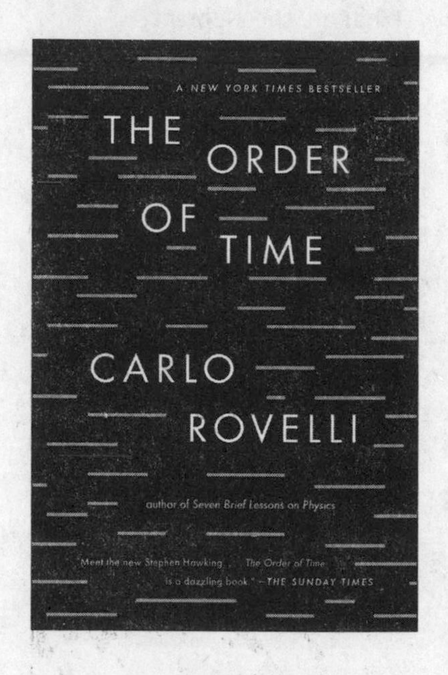

One of *Time*'s Ten Best Nonfiction Books of the Decade

"Meet the new Stephen Hawking. . . . *The Order of Time* is a dazzling book."

—*The Sunday Times* (London)

"An elegant grapple with one of physics' deepest mysteries."

—The Wall Street Journal

"Rovelli's new story of time is elegant and lucidly told."

—The Washington Post